Acknowledgement

I owe gratitude to my teachers, who taught me, at an early age, how to pay attention to details and seek multiple views on any one matter. I owe it to those who insisted on letting me learn and practice the craft of writing and the act of sharing. The people we meet and confer with every day are great sources of information, knowledge and wisdom. I owe special thanks to friends and family who encouraged me to put my thoughts together in writing. I thank those who have shared books with me on various occasions, and grateful to reach this one day, where those books would help write what you are about to read.

I submit my special thanks to my daughter Lina and my son Iosif for helping me to find and retrieve specific sources of geographic information and to edit the final version of the book. Their continuous encouragement and zeal for the subject have given me the wind I needed to set sail. I thank in advance those of you who shall read and reflect on the ideas brought forward in this book. I believe your reflection, comments, sharing and/ or actions will greatly improve our lives and jump-start the gain ahead of the pain.

Preface

Numerous pieces of information from history, geology, biology, astronomy, and science have come to my attention at random over the past decade. Those included a variety of books that my family and friends have given me, documentaries displayed on TV, history citations from holy books that started to appear clearer, my personal and academic background in physics and electronics; all have offered me pieces of a jigsaw puzzle. When I began to reflect and build the pieces together, I became aware of the big picture and therefore had the urge to write and share with others.

Several questions emerged in the most recent years but no one could still give definite answers; why has the annual number of Earthquakes, above 6 Richter, increased more than double starting 2007? Where is the thermal energy, which forms the Temperature Belts of homogenous climates on the surface of Earth, coming from? Why is North America getting warmer and shorter winters while Europe is getting colder and longer ones? Is there a relation between the weakening of Earth magnetic field and the increase in Global Warming? Could the reasons behind Earth Ice Cap melting be the same reasons behind Mars Ice Cap receding? Why is the Ice growing in East Antarctica? Why are there patches of Reversed magnetic fields in Antarctica? What makes Earth spin about its axis, speed up at times and slow down at other times by milliseconds a day? Will the Earth slow down its spin speed, similar to what Venus and Saturn have experienced in recent years? What is the cycle of Climate Change? How did ancient text, in holy books, speak about several Climate Changes? What wrong did historians do when they decoded the Mayan Calendar? And why December 2012 date, for an end of cycle and a beginning of a new one, is inaccurate? And what is the correct date?

MORE THAN 60 MINUTES

When Earth Stands Still

T.S. NIAZI

Copyright © 2009-2012 T.S. Niazi
All rights reserved.

ISBN-13: 978-1478362814
ISBN-10: 1478362812

An important part of this book is designed to draw public attention to discoveries made by scientists and explorers around the world concerning our planet, its state and its development. Without the underpinning work of those scientists, there would be a little foundation to base camp upon. I have attempted to present their works in their own words whenever possible. The integration, with a different insight of such discoveries, existing knowledge, history citations in holy nooks and ancient texts have produced a completely new reasoning behind Earth changes. Although most, if not all, of the puzzle components of Earth changes are based on established facts and theories, further due diligence of some of the planetary, historical and/or human discoveries could still be required.

Most, if not all of independent scientific reports, documentaries and global news networks link climate change and global warming to human malpractice with nature, increased industrial and urban waste, deforestation and increased carbon emission. I will shed light through a different angle and demonstrate that this may not be totally true. Most text books explain that temperature, at sea level, is highest at the geographic Equator and lowest at the Poles. I will explain that it could be true as long as the magnetic poles are closely aligned with the geographic poles. However, when alignment is no longer in place, the meteorological map of the planet will tilt, following the travel of the magnetic poles. While the basics of magnetism, the rising frequency of earthquakes, the characteristics of the geological layers of Earth, and temperature variation may explain what is happening today, the ancient texts of past civilizations and history records that are imbedded in holy books, on the other hand, paint a cycle of repetition of Earth changes.

The Source chapter explores the human migration as a response to Earth changes. A significant event that occurred some seventy thousand years ago compelled the commencement of mankind's migration out of Africa. But, that is not all, as the human brain picked up an amazing pace of development since the journey out of Africa had begun. Upon examining the human brain and exploring its prescient capabilities, we find ourselves channeled back into the vast universe. Quantum physics, or the science that explains the subatomic particles behavior, could explain why there are times when some people can invent and innovate while others can not. Some scientists believe that energy and information are two sides of the same coin. As the universe expands and its energy is conserved but is mutating in different forms, could energy levels be decoded as chunks of information? And could the human brain be equipped with a capability to receive and transmit energies and therefore information within the vast universe around us? Could this capability bring us closer to a *"single theory for everything"*? And provoke all kinds of discoveries? A friend of mine asked me if such brain capability helped me to connect to the source of information that led, through cause-and-effect, to understand Earth changes and manifest the cycle of repetition.

The Earth chapter provides basic information on the planet's main characteristics including shape, geological structure, material composition, magnetic stealth and atmosphere. It highlights the relation between its magnetic field and the amount of thermal energy received on the surface of the Earth. Most scientists attribute the Earth spin about its axis to the planet angular momentum that emerged when the Earth was initially formed billions of years ago. They claim that if it stops, it will not spin again! However, they could not explain why the Earth's spin about its axis speeds up and slows down by some milliseconds a day. Most scientists claim that the variation of temperature on the surface of the planet is dependent on the thickness of the Ozone layer. They fail to explain why, when the Ozone layer thickness varies, the planet surface beneath remains about the same temperature. Others attribute the temperature belts of homogenous climate zones to the 'angle of projection' of the solar energy onto the surface of the planet. They overlook the fact that the Earth is tilted by a constant 23.44° angle about the axis perpendicular to its orbit around the Sun. For instance, the angle of projection is almost the same at both the Tropical Belt and the northern Middle Latitude Belt on June 21, yet the temperatures vary greatly between the two regions. Most scientists attribute the Outer Core, the low viscosity liquid layer of the planet, to be the source of the magnetic field that we observe on the surface of the Earth. They overlooked the fact that if such a magnetic field is induced by the electrons flowing in the Outer Core layer (moving from the Inner Core towards the inner Mantle), then the orientation of such induced magnetic field would be perfectly parallel to the spin axis of the planet. In other words, the magnetic pole and the geographic pole would be aligned on exactly the same geographical coordinates. Yet, we know that the magnetic poles are constantly wandering far and around the geographic poles. The two magnetic poles wander independently from one another and are not at directly opposite positions on the planet. I try to address the abovementioned ambiguities by leveraging the basic concepts and theories of electricity, magnetism and laws of motion, I try to address some basic questions, for instance: Why the orientation, rather than the angle (23.44°), of the Earth's axis, is believed to change over time, completing a full circle in 25,800-years cycle? What makes the magnetic poles wander beyond and around the geographic poles? What causes the spin of the Earth about its axis to speed up at times and to slow down at other times? Would the time come when the spin becomes very slow and an hour of time lasts *More Than 60 Minutes*? And what protects the Earth from the solar energy? Why is the temperature hottest at the Equator and coldest at the poles when measured at sea level? The spinning of the Earth creates the equatorial bulge. The equatorial diameter is 43 km (27 miles) larger than the geo pole-to-pole diameter. Could other geological layers of the planet be also bulged? And how acute is the bulge angle? What makes the Earth spin counterclockwise? Could such a spin be brought to a standstill? Could the spinning direction be reversed? Would the Sun, at any given point in time, rise from the West and set in the East? By addressing the above questions in this book, I challenge several commonly agreed hypotheses that are made available by scientists, astronauts and others about Earth. Connecting geological discoveries, a scenario of new locations of the magnetic poles, is explored and the results are investigated including increased rate of major quakes, volcanoes and changes of the temperature and the precipitation map. Life on the planet is responding in various ways.

The Cycle chapter takes us back through history at times when the Earth and its climate experienced similar change syndromes as the one that it is passing through today. In an effort to locate the cause and frequency, I explain how the emergence of an object in the sky influenced and assisted Moses Exodus around 1,550 BC and four decades later assisted Joshua at his wars when the Sun and the Moon, or rather the Earth stood still. More than 3,000 years earlier Abraham saw a planet in the sky and was wondering if it were the creator of the world. How could Abraham have witnessed a planet in the sky with bare eyes?! There was no space telescope at the time. Was there a planet or a planetoid that was cruising close enough from Earth to be as visible as the Moon or the Sun? Historians claim Abraham to have predated Moses by 500 years only. Discoveries from Ancient Egypt proved otherwise. Some 10,600 years ago, in anticipation of the Earth changes, Noah built the Ark that helped him, family and some of the living beings, at the time, to survive a tidal wave and an epic flood. A cause-and-effect of a 3,562-year cataclysmic cycle is explained. We seem to have just started to touch the periphery of the end of one cycle which is followed by another in the not too distant future. Could we correct the inaccurately decoded Maya Calendar date of December 2012?

The Mandate chapter emphasizes the early anticipation and correct strategies to be adopted in selected industries. Instead of spending time and resources on combating increased carbon emission, a strategic plan to a smooth transition to an exchanged climate becomes more critical than ever, if the climax of Earth changes is reached in 2017. The early anticipation and correct strategies should be adopted in selected industries in order to i) minimize the loss of resources, ii) maximize the protection of species and iii) facilitate our recovery and transformation into different climatic, social and awakening conditions. If there will be a change in the global precipitation map, do we have enough stores of grains and transportation means to avoid global famine? What could be done to ensure cooperation between countries that will be experiencing a major agriculture shift and how fast can the transformation go? More elaboration and recovery approaches are discussed over disciplines such as finance, infrastructure, energy, plant life and civil order.

My academic background is in electronics and physics. My experience is in informatics, investment, development and entrepreneurship. Due to the urgency of the matter, the book I authored and published, may not be an elaborate and detailed piece of work in which all aspects are thoroughly drilled and neatly woven together. However, I believe that it is prudent that experts in other fields can contribute deeper on parallel analyses, whether on a new planet approaching our inner solar space or a vulnerable Inner Core that is very sensitive to magnetic fluctuation or a whole thread of cause-and-effect relations that connect all changes. I expect this book to bring a considerable controversy amongst scientists and interested scholars as it links geological, historical, and astronomical records as well as theories of physics and trigonometry to reach conclusive evidence of the geophysics that govern our planet and make it spin about its axis; forming the Temperature Belts, experiencing Global Warming and Climate Change as well as encountering an alarming increase of Earthquakes and Volcanoes. I expect to be met at first with opposition from orthodox scholars. This, unfortunately, is a common response of orthodox historians, scientists and geologists towards new ideas, especially when these emanate from unqualified sources. It is however self-defeating, for in the end, new ideas and integration of existing theories with a different insight will always triumph over the old.

We come as one people on a continuously changing space ship that we call Earth. The changes are accelerating into epic proportions and we can do nothing to stop them. We should capitalize on the opportunity and plan to be prepared before it is too late. As is always said, "Failing to plan is planning to fail".

Table of Contents

Chapter 1: The Source 1
    Remote viewing 1
    The brain 5
    Quantum Physics, ... 8
    The Genographic Project 15
    Theory of Everything 23
    Revelation 26
Chapter 2: The Earth 29
    Strange sky sounds. 30
    Sudden death of birds and fish. 32
    Aluminum versus carbon composites. 33
    Climate and pole changes. 34
    Navigation 35
    Wandering poles 36
    Pole reversal 39
    Why Earth temperature varies at sea level? 40
    The Earth's structure 50
    The atmospheric pressure. 52
    Water vapour 53
    The magnetic field 55
    What makes Earth spin 60
    Position and angle variations 66
    The tilting pole 68
    A new route is breaking the ice 73
    Mars temperature. 74
    Mars magnets setup. 77
    Mars global warming. 79
    Mars spinning speed. 80

Chapter 3: The Cycle 81
The Exodus mystery 81
Asteroid impact 86
Abraham saw a planet. 88
When Earth stopped 93
What about Noah, ... 95
The Moon's a balloon, ... 98
The Moon splits 100
"The Hour" is mentioned 101
The Mayan calendar 102
Other observations from earlier civilizations ... 104
Cycle features 106
The twelfth planet 108

Chapter 4: The Mandate 113
Capital markets 113
Managing wealth 114
Agriculture 117
Infrastructure 118
Energy 121
Planet life 126
Civil order 129
Lead country projects, ... 130

Bibliography 134

The Source

Remote viewing refers to the attempt to gather information about a distant or unseen target using paranormal means or extra-sensory perception. The term was introduced by parapsychologists Russell Targ and Harold Puthoff in 1974 during their work at the Stanford Research Institute (SRI). It is defined as the acquisition and description, by mental means, of information blocked from ordinary perception by distance, shielding or time. It was announced in the 1990s, following the declassification of documents related to the Stargate Project, a research program sponsored by the USA federal government, to determine any potential military application of psychic phenomena. The program was terminated in 1995, citing a lack of documented evidence that it had any value to the intelligence community.

Remote viewing has been used to discover mineral deposits, hidden treasures and missing people (Targ, 2004). In his book the *Limitless Mind,* Targ explains how the kidnapping of Patricia Hearst was resolved using remote viewing. In February 1974, a group of terrorists kidnapped the nineteen-year-old American newspaper legendary William Randolph Hearst's heiress named Patricia from her apartment near the University of California at Berkeley, where she was a student. Days after the kidnapping, the police still remained clueless. It was such a desperate situation that the Berkeley police department thought about seeking psychic guidance. They called on the SRI. Pat Price, a remote viewer who was a retired police commissioner from Burbank, because he mentioned that he had often worked on similar situations. At the police station, the detectives had a lot of questions unanswered. However, Price asked the detective working on the case for a 'mug book' of pictures of local people who were recently released from prison. Price took the book and laid it flat on a table and looked through four mug shots on each page. Then, at about ten pages later into the book, he placed his finger on one of the pictures and said, "He's the leader". The man who Price singled out from the mug book named Donald Cinque DeFreeze, had managed to escape from California's Soledad Prison a year earlier. The Police had no idea where to find DeFreeze. So they asked Price if he could determine where DeFreeze might have gone. In a moment of silence, Price visualized a white station wagon to the north parked on the side of the road where the kidnapping had taken place. However, he also visualized that it had just been abandoned. Upon sharing his visualization and location with the police, they located the abandoned white van in the vicinity of Price's description. With the help of Price's more visualizations, the police was able to locate the hideout and free the kidnapped girl.

Joseph McMoneagle, America's most decorated and world-renowned remote viewer, served in the Stargate program from its beginning to its termination in 1995. In 1970, while serving overseas in the Army, Joseph McMoneagle had a near-death experience and claims that it has opened a channel in his brain to spontaneous psychic flashes. After this incident, he was able to perceive and describe previously unknown people, places, objects, or events more accurately. Later, when tested at SRI International, he was confirmed as a world-class remote viewer. McMoneagle suggests an explanation for the way the Pyramids at Giza were built (McMoneagle, 1998).He suggests that water was effectively used to aid in transportation of building blocks. The builders had large blocks of rock cut a few kilometers away from the construction site, and placed them on sleds then moved down to docks, loaded on rafts and shipped to the final site which was already submerged under water of manmade lake. The first few rows of stone were laid perfectly flat, fitted exactly and with great precision. Not only was water used to lubricate the saws, but it was also used to provide a perfectly horizontal level on which each of the stones was exactly cut. The water also had some softening effect on the stone, making it more malleable to the tools used.

McMoneagle had noticed, through remote viewing, that the water was brought through a series of locks between the river and the construction site. This canal led to making an artificial lake at the construction site which was surrounded by a retaining wall. McMoneagle noticed some interesting cycles occurring within the lake. Its size and depth were increasing over time. Such controlled flooding was accomplished through the annual addition of more stone to the wall that engulfed the artificial lake. As the level of water increased every year, so did the construction level. Ancient Egyptians documented the period it took to build the Great Pyramid. They engraved, in petroglyph, that it had taken 'twenty floods' to complete the Cheops Pyramid. I was always wondering why they did not scribe twenty years instead!

The volume of the Nile's flood varies from one year to another. At low floods, when the water was not leveling up to the desired height of the construction, the work was put on hold. In his book, *the Ultimate Time Machine*, Joseph McMoneagle provides an estimate of a "fifty years" period it had taken to complete the construction. A shallow version of the walled lake, described within the remote viewing session, might have been sufficient to lay-in the first few of layers of base blocks, ensuring success in the nearly impossible task of beginning with a near-perfect foundation plane. "From an engineering standpoint it is nearly impossible to achieve a perfect flat and balanced plane in the first few rows of stone", wrote McMoneagle. Attempts to replicate pyramid construction using modern construction techniques and equipment have failed because of the inability to produce this very exact, near-perfect plane of reference at the base. It is only achievable through unmistakably flat, perfectly still water. If one does not begin with a near-perfect base, the difference in weight from corner to corner tends to cause a pyramid to self-destruct long before it is finished. Using rafts to transport large stones and replacing wheels with sleds makes perfect sense. In his book, McMoneagle explains that the area surrounding the Great Pyramid, which is now arid and dry, was most certainly a rainforest or jungle area.

In his BBC documentary (Cruickshank, 2007) Dan Cruickshank travels the world in search of mankind's greatest creations. In his trip to Upper Egypt, he visited the Valley of the Kings, which began with the eighteenth dynasty and ended with the twentieth. The kings abandoned the Memphis area and built their own tombs in Thebes. They had also abandoned the pyramid-style tombs. The valley, where they built the tombs, forms the backdrop to the temple complex of El-Deir El-Bahari. The Valley of the Kings is hidden from sight, behind the cliffs. Although the most direct route to the valley from the Nile is a rather steep climb over these cliffs, a much longer, shallower route existed along the bottom of the valley. This was quite possibly used by funeral processions, pulling funeral equipment by sledges to the rock-cut tombs in the valley. Cruickshank noticed that it was possible to reach the valley from the workers'

Figure 1- Pharaonic Egypt

village in about thirty minutes by walking along the steep mountain paths. He wondered why they chose to live in a place six kilometers away from the Nile; the only source of water in current days. It could be that at the time the workers lived there, the whole terrain was lush forest and that water was abundant, much like that which Joe McMoneagle had observed in his remote viewing session! Recent research by University College London (UCL), over the North Africa aquifers, dates its waters to some 7,000 years ago. It is deduced that at that time, the region was lush green, full of fresh water lakes and plant and animal life. Due to sudden climate change, inhabitants had to depart in a hurry. Several artifacts were found in caves in the Libyan desert, according to Professor Dr. Hassan Fekri, Petrie Professor of Archaeology at the Institute of Archaeology & Department of Egyptology, UCL (1994- 2008). But, how does the brain function to acquire and manifest such prescient capabilities of remote viewing?

The brain is an electrochemical organ using electromagnetic energy to function. Electrical activity emanating from the brain is displayed in the form of brainwaves. There are four categories of brainwaves, ranging from high amplitude and low frequency to low amplitude and high frequency. Men, women and children of all ages experience the same characteristic brainwaves. They are consistent across cultures and national boundaries; Research has confirmed the beneficial aspects of meditation (University-of-California-Los-Angeles, 2009). In addition to having better focus and control over their emotions, many people who meditate regularly have reduced levels of stress and strengthened the immune system. People who meditate grow larger brains than those who don't. Researchers at Harvard, Yale, and MIT (Lazar, Treadway, & Chakrapami, 2006) have found the first evidence that meditation can alter the physical structure of our brains. Brain scans reveal that experienced meditators boast increased thickness in parts of the brain that deal with attention and processing sensory input. In one area of gray matter, the thickening turns out to be more pronounced in older people than in younger ones. It is intriguing because those sections of the human cortex, or thinking cap, normally get thinner as we age. The goal of meditation is to pay attention to sensory experience, rather than to our thoughts about the sensory experience. When you suddenly hear a noise, you don't analyze it but rather just hear it. If there is absolutely no noise, you pay attention to your breathing. Successful meditators get used to not thinking or elaborating things in their minds. During meditation, people often feel no sense of the space surrounding them. Scientists investigating the effect of the meditative state on Buddhist monks' brains have found that portions of the organ previously active becomes calm, whilst inactive areas become stimulated. Using a brain imaging technique, Dr. Newberg, a radiologist at the University of Pennsylvania, US, and his team (Newberg, 2002) studied a group of Tibetan Buddhist monks as they meditated for approximately one hour. The monks were asked, that when they reached a transcendental high, to pull a kite string to their right to release an injection of a radioactive tracer. By injecting a tiny amount of radioactive marker into the bloodstream of a deep meditator, the scientists soon saw how the dye moved to active parts of the brain. Later, once the subjects had finished meditating, the regions that were imaged of the meditation state were compared to the normal waking state. The scans provided remarkable clues about what goes on in the brain during meditation.

"There was an increase in activity in the front part of the brain, the area that is activated when anyone focuses attention on a particular task", explained Newberg. In addition, a notable decrease in activity in the back part of the brain, or 'parietal lobe', recognized as the area responsible for orientation, reinforced the general suggestion that meditation leads to a lack of spatial awareness. This explains the fact that while in the waking state, we experience three-dimensional space axes and a single dimensional time axis, at deep meditation or at sleeping stage, space axes are reduced to nil while time axes became more than one, enabling conscious to move into past and future frames of events. Would the brain, at such moments, function as a receiver and capture the vision and/ or knowledge stored at diverse time and space? The answer is offered by the pineal gland, which is called the 'third eye' by some. It is a small endocrine gland in the vertebrate brain that produces the melatonin hormone that affects the modulation of wake/sleep patterns & photoperiodic functions (Macchi & Bruce, 2004).

It is shaped like a tiny pinecone; hence its name, and is located near the center of the brain, between the two hemispheres, tucked in a groove where the two rounded thalamic bodies join. It is part of the epithalamus, a midline structure,

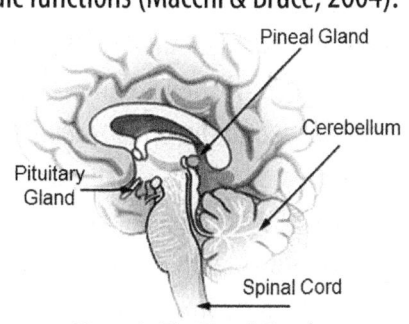

Figure 2- The Pineal Gland

and is often seen in plain skull X-rays, as it is often calcified. Calcification is typically due to intake of the fluoride found in water and toothpaste. It was the last endocrine gland to have its function discovered. The pineal gland is activated by light and it controls the various bio-rhythms of the body. It works in harmony with the hypothalamus gland, which directs body's thirst, hunger, sexual desire and the biological clock that determines our aging process. The physiological function of the pineal gland has been unknown until recent times. Mystical traditions and esoteric schools have long known this area in the middle of the brain to be the connecting link between the physical and spiritual worlds. Considered the most powerful and highest source of ethereal energy available to humans, the pineal gland has always been important in initiating supernatural powers. Development of psychic talents has been closely associated with this organ of higher vision. To activate the 'third eye' is to raise one's awareness frequency and move into higher consciousness that is perceived through the 'eye of time' or 'third eye'. Meditation, visualization, yoga, and all forms of out of body travel open the third eye and allow one to view beyond the physical. With practice, these experiences become faster and more frequent. Psychic abilities increase as well as dream-time messages. One may first begin with one's eyes closed, but with practice, the third eye would be fully opened by focusing one's attention and receiving messages with one's physical eyes open. However, the thread of knowledge or information downloaded to the brain through the pineal gland encompasses more than imagery. It also includes the associated senses or information around it, such as smell, sound, taste and touch. Nonetheless, there remains a major challenge in finding out the mechanism of information storage, retrieval and interpretation by the human brain. We needed the helping hands of science to go deeper than the atomic level; and that science is called Quantum Physics.

Quantum Physics, which is also known as Quantum Mechanics, can help us understand how knowledge is stored in the surrounding universe. How could we listen to and learn from it? And how could we store and deposit our experiences and thoughts in it? So, let me start by sharing some background. The world is composed of elementary particles: electrons, protons, photons, phonons and so on. In the early days of the twentieth century, the Danish physicist Niels Bohr introduced a principle called wave- particle duality. This principle refers to the fact that the phenomena we normally think of as waves, such as sound or light, are actually made of particles, or quanta (quantum is a Latin word for *how much*). Particles of light are called photons, and particles of sound are called phonons. Alternatively, the phenomena that we think of as particles, such as electrons or atoms, have waves associated about their positions. The particle is more likely to be found in places where the wave is large. The double-slit experiment, as shown in Figure 3, demonstrates the wave nature of the particles. In the experiment, the first wall has one slit in front of the light source. Therefore, it is expected that the light beam will pass through the slit and keep travelling at a straight line. The second wall has two slits that are neither aligned with the slit of the first wall, nor the beam of light that travels through it. One may not expect to see any light passing through the double-slits of the second wall, but we find alternating bands of light and darkness to appear on the screen behind the second wall. It turns out that as the beam of light alternates between particle and wave, the beam goes through the single slit of the first wall and the double-slits of the second wall before hitting the screen at the back. You may observe 'alternating bands of light', where the peaks and troughs of the wave from one slit coincide with the peaks and troughs from the other slit and get reinforced. You also get to observe 'alternating bands of darkness' where the peaks and troughs of the wave from one slit coincide with

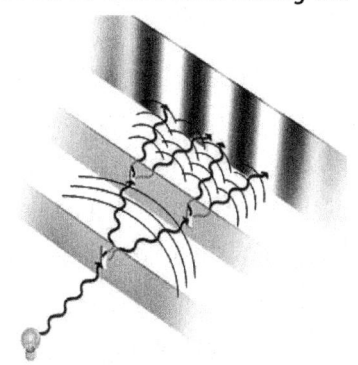

Figure 3- Wave like nature of the light particles

the peaks and troughs from the other slit and cancel each other out. The pattern that the particles make on a screen exhibits an 'interference pattern', evidence for the underlying wavelike nature of the light particles. The double-slit experiment illustrates the fact that a particle doesn't have to be either in one place or the other. Because of its associated wavelike nature, a sub atomic particle can be both 'here' and 'there' at the same time.

This ability of things to be in many places at once is responsible for the power of quantum computation. But, "If things can be in two places at same time, then why don't we see pebbles, people and planets showing up in more places than one?" Dr. Seth Lloyd, Professor of mechanical engineering at MIT, asks. The bigger an object is, the greater its tendency to behave more 'classically' and less 'quantum-mechanically' (Lloyd, 2005). The reason does not lie so much with the physical size of the object but with its visibility. The bigger an object is the more interactions it will tend to have with its surroundings, which makes it easier to detect. For a particle to go through both slits at once and produce an interference pattern, in the double-slit experiment, it must pass through the slits undetected, meaning without interaction with anything else.

Since each particle in nature may oscillate between many energy levels, could each elementary piece of a physical system register a chunk of information? To keep it simple let us limit our discussion in the energy levels of a particle on these pages to two levels only. A particle oscillation could be regarded as '1' or '0', 'on' or 'off', 'high' or 'low'. Combination of such binary representation (also called **bi**nary digi**ts** or bits) could be infinite depending on the number of bits. For instance in an 8-bit frame, the binary code 1111 0001 is an eight-bit combination that, we can agree, will represent the decimal number '1'. Or the binary code 1110 0010 is an eight-bit combination that, we can agree, will represent the alpha letter 'S'. We can set such code as the standard code amongst all computers; if we so wished. Since the dawn of digital computing, the different manufacturers have agreed to follow standard tables for character encoding, storing, retrieval, processing, decoding and interchange. If, for instance, the Extended Binary Coded Decimal Interchange Code (EBCDIC), which is an eight-bit character encoding (code page) that is used on most IBM mainframe operating systems, as well as IBM midrange computer operating systems, was used to decode 0101 1011 in a character that we understand, it would be the exclamation point '!'. However, if the extended American Standard Code for Information Interchange (extended ASCII), which is also an eight-bit character encoding that is used by Windows, UNIX and Linux systems, was applied to decode the same string 0101 1011 in a character that we understand, it would be the uppercase letter 'Z'. So while the information string is the same, it depends on the code page that is used by the recipient to decipher it according to the intent of the one who created and stored it. Binary code was adopted by computer manufacturers to fit the limitation of underlying transistor based double-state circuitry; 'on' or 'off', 'high' or 'low', '1' or '0'. Energy levels in any one atom are more intricate and certainly can exist in more than 2 energy level states. Therefore, it is suggested that a single atom will contain more information than only 2. Instead of the Binary code; the Octal, Decimal, Hexadecimal or bigger code page could be used, depending on the maximum energy levels in any one atom.

In 1993, Lloyd discovered a way to build a quantum computer. Quantum Physics is the theory that describes how matter and energy behave at their most fundamental sub-atomic level. On a smaller scale, Quantum Physics describes the behavior of molecules, atoms, and elementary particles. On a larger scale, it describes the behaviour of living things. On an even larger scale, it describes the behavior of the universe as a whole. The universe is the biggest thing there is, and the 'bit' is the smallest possible chunk of information. "The universe is made of bits", wrote Lloyd. Every interaction among those pieces of the universe processes information by altering bits. The universe computes, in an essentially Quantum Physics fashion, because the universe is governed by the laws of Quantum Physics. Its quantum bits are called qubits. The history of the universe is a huge and an ongoing quantum computation. Simply put, the universe is the largest quantum computer ever.

As soon as the universe began, it began computing. Until the formation of atoms, almost all of the information in the universe lay at the level of the elementary particle. Nearly all qubits were registered by the positions and velocities of protons, electrons and so forth. On any large scale, the universe still contained very little information. It was featureless and close to uniform. In Quantum Physics, quantities such as position, velocity and energy density do not have exact values. Their values are in constant fluctuation. Probable values could be described, but not with perfect certainty. As time passed, the attractive forces of gravity caused more matter to move toward denser regions, thus further increasing their energy density and decreasing energy density in the surrounding volume. Lloyd states that Quantum Physics describes energy in terms of quantum fields, whose weave makes up the elementary particles, which are protons, electrons, and quarks. The energy that we observe around us, in the form of the Earth, stars, light, and heat, was drawn out of the quantum fields by the expansion of the universe. As the universe expands, gravity sucks energy out of the quantum fields where the energy is always almost positive and is exactly balanced with the negative energy of the gravitational attraction. As the universe keeps expanding, more and more positive energy becomes available, in the form of matter and light and is balanced by the negative energy in the attractive force of the gravitational field. The more energy, the faster the atomic particle flips between its levels and, therefore, the ability of the qubit to represent more information.

Ultimately, information and energy play complementary roles in the universe. "Energy makes physical systems do things. Information tells them what to do", explained Lloyd. Not only can atoms emit light, but they can also absorb it. Just as an atom can jump from a higher energy level to a lower one, emitting a photon in the process, an atom can absorb a photon and can jump from a lower level of energy into a higher one. Take an atom in its ground level and bathe it with a beam of laser light made up of photons or a magnetic field that is also made of photons. If bathing photon energy equals the difference in energy between the ground level of the atom and its next lowest energy level, the atom will absorb a photon from the beam and jump from the ground level to the next level. If the energy is less than the difference in energy between the ground level of the atom and its next lowest energy level, the atom will not absorb a photon from the beam and will remain at its ground level.

Core rings were used to store data and programs in early digital computers. If a ring is magnetized, it indicates an 'on', 'high' or '1' condition. If it is not magnetized, it indicates an 'off, 'low' or '0' condition. Attempts to sense if a core ring is magnetized will make the core ring lose its magnetic charge. The reading of information in such a case should be followed by a re-write or re-charge back of the core ring as it was, hence, the information is kept intact. Would the same process take place as an atom energy state is measured? Or would dual wave-particle property bring a particle, or a qubit, in two places at same time so no loss of energy status of the original particle is observed? There remains the decoding process or algorithm to decipher such combinations of atomic energy states into meaningful information. Series of qubits could represent music, images, texts, formulae, and so on.

The bridge or channel that could be established between the information sources as stored in energy levels of atoms, and the pineal gland of a human brain remains to be enlightened. "When you eat or drink any substance that contains carbon atoms, the digestive system secretes enzymes that convert the sugar of the edible substance into glucose, a form of sugar that can be used by the muscles to generate energy function as intended", wrote Lloyd. Depending on the effort that is expended by the muscles, energy is converted into thermal energy and work. Even when the body is at a standstill, there is a minimal conversion going on to keep the muscles ready for action. Therefore, it is expected, that several atoms of the body are oscillating and changing from ground level to the next lowest energy level. Such a process is cyclical, so that when the muscle heats up it pulls photons to help atoms move into the next lowest possible energy state. The reverse process takes place when the muscle cools off, then getting rid of extra photons. Most of these photons constitute the magnetic field or aura, a field of luminous radiation around a person or object. There is an interchange going on, because the human body lives inside a universe of energy and information. The human magnetic field is engaged in a function similar to that of the magnetosphere that surrounds the Earth and limits the penetration of solar charged particles and cosmic rays. In daylight, many of the magnetic field photons are therefore outbound and busy shielding the body. Whereas, when the Sun sets many of the magnetic field photons are available to assist the exchange information that is stored in the structures of nature around us. It comes as no surprise that the best time for meditation is in the late evening hours. It also comes as no surprise that body metabolism is highly active during daylight when the need to protect the body from solar energy and cosmic rays is greatest, as opposed to night time. As explained earlier, the pineal gland acts as the portal through which the information is exchanged. However, how could a sequence of data streams that is downloaded through the pineal gland be interpreted to something meaningful to the human brain? A computer industry standard called XML can help us imagine how the downloaded qubits of information could be interpreted. XML stands for 'Extensible Markup Language' and is created to help data sharing across multiple systems with different coding/decoding backgrounds. The concept is simple: data is always stored with a header that explains its intent. Is it a 'date', a 'weight', a 'name', an 'address' or else? And what sort of code page is being used to create it? A list of semantic constraints is developed by standard agencies, so that data ambiguity no longer exists between source creator and target recipient of a stream of information. It would be ideal, if, while the universe is expanding upon energy and information, there could be information about the information so that the right interpretation is applied and a true vision is obtained by the brain whether we are awake or asleep. If this is the case, will the capability be different from one society to another, or one generation to the other? And, how much are humans different from one another or from one era to the other? If geological and historical records are scarce, could Genetic research help establish the evolution of human brain capability along the migration to populate the planet over the past 70,000 years?

The Genographic Project is a collaborative effort between National Geographic and IBM to chart new knowledge about the migratory history of the human species by using sophisticated laboratory and computer analysis of DNA of hundreds of thousands of people from around the world. DNA, or deoxyribonucleic acid, is the hereditary material in humans and almost all other organisms. In this unprecedented and contemporary research effort, the Genographic Project is closing the gaps in what science knows today about humankind's ancient migration stories. The Genographic Project is a five-year research partnership led by National Geographic explorer Dr. Spencer Wells. A team of renowned international scientists and IBM researchers headed by Wells are using advanced genetic and computational technologies to analyze historical patterns in DNA from participants around the world to better understand our human genetic roots. The team gathers field research data in collaboration with indigenous and traditional peoples around the world. Scientific evidence suggests that the human species ultimately traces back to Africa. However, how did our ancestors migrate and populate the world? The Genographic project aims to uncover some of these mysteries and discover the details of why and when did humankind migrate around the world. What impact did culture have on the human genetic variation and how did cultural practices affect patterns of genetic diversity? If we share a recent common ancestry, why do we look different from each other? The human body is made of some 50 to 100 trillion cells. Each forms a unit of life and combines to form more complex tissues and organs. Inside each cell, genes comprise a blueprint for protein production that determines how the cell will function. The double-helix-shaped molecule that holds an organism's genetic information is called DNA or Deoxyribonucleic Acid. Genes determine the physical characteristics and the complete set of some 30,000 to 40,000 genes is called the genome. The genetic structures, that are called chromosomes, carry heredity. Humans have twenty-two numbered pairs of chromosomes and single pair of sex chromosomes which is represented as XX in female and XY in males. Each chromosomal pair includes one inherited from the father and one from the mother. While the Y chromosome traces the line of male through his ancestry, random mutations in the DNA sequence can occur. Small and in bred groups of any species are subject to rapid genetic divergence because of disruption of natural selection and genetic drift. Such random changes in the DNA sequence are called 'markers'. Once markers have been identified they can be traced back in time to their origin—the earliest common ancestor of anyone who carries the marker. It was therefore possible to trace the human migration path through tracing the origin of sampled DNA back in time. No one knows when modern humans first appeared on Earth.

Paleoanthropologist Richard Leakey found two skulls and various bones of modern humans between 1967 and 1974. They have been dated to be about 195,000 years old. On the South African coast there exists a warren of cave openings and an overhang that is called the Klasies River Cave that is home to an array of artifacts from the Middle Stone Age, i.e. about 60,000 to 130,000 years ago. Tools made of nonlocal materials suggest that they were used for trade, which is considered a very modern culture behavior for the period of the Middle Stone Age. The Middle Stone Age, Lake Toba, in Sumatra, was the scene of the largest volcanic eruption of the last two million years, a blast that is nearly 3,000 times larger than the 1980 eruption of Mount St. Helens. Following the 73,000 BC eruption, worldwide temperatures may have dropped significantly causing a period of volcanic winter. Some anthropologists believe that the frigid climate and drought wiped out many of Earth's hominids, leaving only small surviving populations. It is not unthinkable that such eruptions and quakes follow galactic influence, as will be discussed in later chapters, and also in cyclic manner, meaning that it could very well happen again, but at various degree of vigor as will be explained in the following chapters of this book. According to the National Geographic website[1], the aftermath of Lake Toba eruption may mark the point in time when the relatively small group that founded the human line of descent began to diverge rapidly and the broad path of genetic diversity emerged in modern humans. The tree of male genetic diversity has, at its root, 'Adam', the common male ancestor of every living man. National Geographic claimed that Adam (meaning *he who has dark skin* in the Arabic language) lived in Africa some 60,000 years ago and that all humans must have lived there at that time.

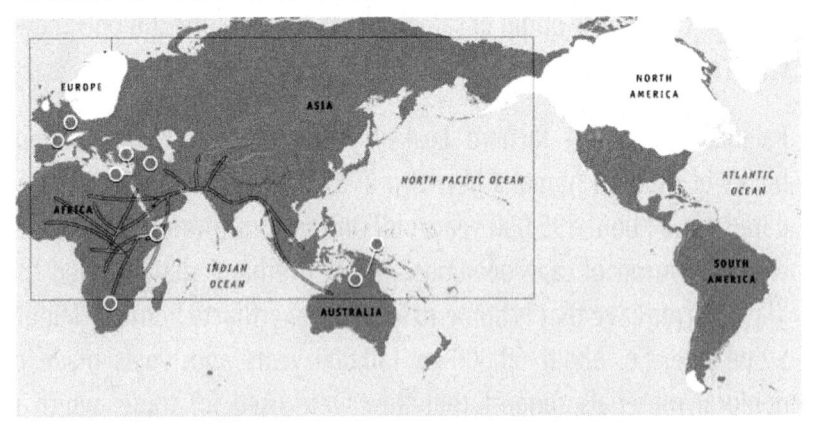

Figure 4- Water level at the sea was lower during the Ice Age

The research suggests that ice had covered more parts of the world than it does now as shown in Figure 4, thus reducing the level of ocean water to a level at which walking paths were made possible connecting Africa to the Arabian Peninsula, Asia to the Americas and Malaysia to Australia. Alternatively, there could be another reason behind lower sea level in the region that lies at the Equator and between the Topics. As Earth spins about its axis, the current speed of earthly material at its Equator is 1,666 km/ hour (1,000 mile/ hour). On the other hand, the speed of earthly material at the Geographic Poles, where the axis of

---

[1] http://www.nationalgeographic.com/genographic

rotation lies is 0 km/hour. Such a difference in speed results in a bulged planet as Earth's crust and oceans at different speeds try to escape to space but the gravitational pull of the planet is restraining them back. If Earth rotated slower about its axis in ancient times, the Oceans bulge would not have existed at the same level as currently witnessed. The sea level between the Tropic of Cancer and Tropic of Capricorn would have been lower. "This Adam was not the only man alive in his era. Rather, he is unique because his descendants are the only ones to survive to the present day", said Wells. In some holy books, while explicit references are made to 'sons of Adam' and other references are made to 'human' in general which clearly highlights a distinction. The hunter-gatherers once known as the Bushmen, in Kenya, carry more ancient evolutionary lineages in their DNA than any other people and exhibit a direct living link to Adam.

Their original pre-historic means of communication is characterized by the clicking sounds that are used for word construction. Such sounds are heard as those used to guide a horse or stimulate dripping drops of water. However, some 60,000 years ago, humans developed the hyoid, which is the tiny, horseshoe-shaped bone that helps in the formation of speech. We still carry some of the early words across many and distant regions of the world. The Inuit, the indigenous Canadian people of the Arctic and Subarctic, drive dogsleds in the winter wilderness and yell 'gee' or 'haa' to direct their dogs to sprint 'left' or 'right'. Remarkably, these two sounds are the same sounds used in the countryside of contemporary Egypt to guide donkeys to move 'left' or 'right'. "In a sense, we are all Africans", said Steven Pinker, a psychologist at the Massachusetts Institute of Technology. He explains that human babies from all over the world have the same basic ability to learn languages, how to count, and how to make and use tools[2]. "It suggests that the distinctively human parts of our intelligence were in place before our ancestors split off into the different continents", says Pinker.

It took humans 10,000 years to reach Australia from Africa, helped by the lower sea level. Arnhem Land Region is one of the five regions of the Northern Territory of Australia. Its stand stone exposures display ancient carvings and paintings of humans hunting and performing ritual acts. "It is estimated that 50,000 to 70,000 years ago human tools became far more refined and diverse", said Wells. Humans began to exploit food resources more efficiently and to create art. A more modern language may have enabled more complex social interactions. The great leap forward delineates a historical point from which its beneficiaries could have migrated, adapted, and settled the Earth. Without such a skill set, the lineage of hominids and humans were destined to extinction. During an ancient era of shifting climates, the Saharan desert gateway in North Africa may have led humans out of the center of Africa and then closed behind them. For Africans, the Ice Age did not mean cold, it meant drought.

Yet some 50,000 years ago, the long-term trend of expanding deserts was interrupted by a period of warmer temperatures and moist climate that made some parts of African Sahara livable. Such a phenomenon is obviously frequent and goes back millions of years at a time where forests flourished for some period of time and then disappear again later, given a change in climate. The disappearance was not the only action that took place. Under pressure and heat, oil was created way

---

[2] http://www.scribd.com/doc/6883119/Steven-Pinker-2001-Evolution-of-the-Mind.

beneath the Earth pointing towards the current oil-basined deep in the Arabian and North African deserts.

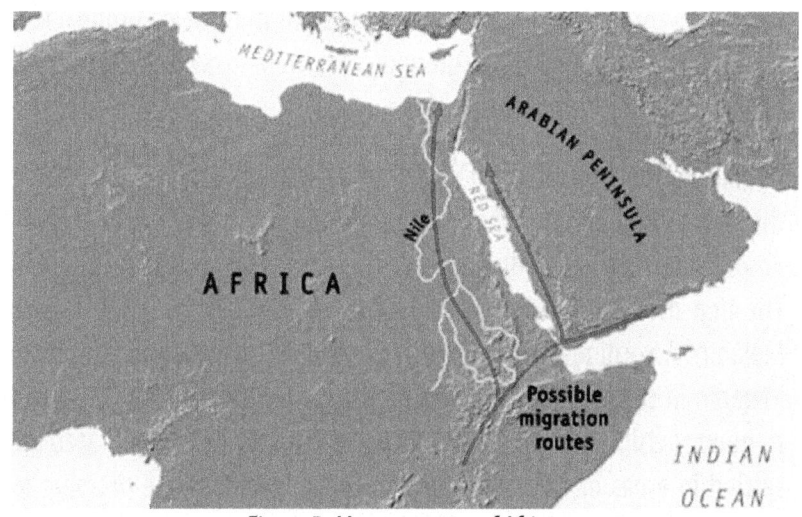

Figure 5- Many ways out of Africa

It is uncertain which route early humans took to reach the Middle East. They may have skirted the Red Sea coast to reach the Sinai Peninsula, traveled north down the Nile to the Mediterranean coast and Sinai, or even crossed the narrow Red Sea strait of Bab Al Mandab to Arabia, given that the sea level was low. Whichever way they traveled out of Africa, the route back was blocked by the later expansion of the Sahara sands. The desert was at its driest between 20,000 to 40,000 years ago. The vast Eurasian and Iranian valleys were unbroken grassland stretching from the Gulf of Aqaba to Mongolia. These valleys were rich with large animals such as antelopes, and in that respect were similar to Africa's savanna.

The environment was ready for upper-Paleolithic-era hunters, who soon began to expand to Central Asia. Thanks to those hunters who dwelled in Central Asia, the rest of the planet was populated. Some of those hunters went over to North India, some to China and Japan, and some crossed the frozen Bering Strait to Alaska and the Americas. Only 10,000 years later they moved on to Europe. 30,000 years ago Europe was going through an Ice Age period. Humans moving in Europe had to take shelter in mountain and ground caves, in lands where the Sun was obscured by thick clouds most of the year.

To counter the harmful Ultra Violet (UV)-radiation arriving from the sky, the human body produces Melanin (an expansive pigment) that absorbs harmful UV-radiation and transforms it into harmless heat. The first modern humans in Africa had relatively large production of Melanin. Accordingly, they had a dark skin. Similarly, the production of Melanin in the iris helps protect the eyes from UV-radiation and high-frequency visible light. As some of these original people migrated and settled in areas of Asia and later in Europe, the selective pressure for Melanin production decreased in climates where radiation from the sky was less intense. The thousands of years of deprivation of sky thermal energy during the ice age in Europe has decreased the production of Melanin and as such turned both skin and iris pigments to lighter colour. Rayleigh scattering is the spreading of light by particles much smaller than the wavelength of the light. It can occur when light travels through transparent solids, liquids and gases. It results from the electric polarization of the particles. The oscillating electric field of a light wave acts on the charges within a particle, causing them to move at the same frequency. The particle therefore becomes a small radiating dipole whose radiation we see as scattered light. Rayleigh scattering of sunlight in the atmosphere diffuses sky radiation, which is the reason for the blue color of the sky. A phenomenon similar to that accounts for the colouring of the eyes in humans, where the pigmentation of the iris varies from light brown to black, depending on the concentration of Melanin, and the cellular density of the stroma. The appearance of blue and green, as well as hazel eyes, results from the Rayleigh scattering of light in the stroma. Neither blue nor green pigments are ever present in the human iris or ocular fluid. Thus eye color follows sustained climate over many generations.

It is occasional that the sole of the foot is directly exposed to sunlight. Due to lack of Melanin production, the original skin pale colour seems preserved at the sole of every human foot, including ethnic Africans.

It is amazing that at certain times along the history line, human brain had gone through quantum leaps; from single 'click' to multi structured 'languages', from adapting to wilderness to harnessing and manipulating of the environment, and from dwelling the Earth to penetrating oceans and space. We are still to understand how human brains have been excited to jump the scale and whether such a brain development is evolutionary, revolutionary or alternating between the two. While some call to modeling the laws of nature into mathematical formulae, others believe that the knowledge acquisition is much simpler than that; as we channel our brain to the universe information source and download answers to complex issues. The question emerges as, which of the two ways, will get faster access to the ultimate knowledge of the single theory that explains it all.

Theory of Everything: was discussed almost two decades ago (Hawking, 1991) "our goal is nothing less than a complete description of the universe we live in", emphasized Stephen Hawking. He thought that there was a good chance that the so-called 'Theory of Everything' would be found before the end of the twentieth century, leaving little for theoretical physicists like him to do. Over the course of the century, scientists were able to explain physical, chemical, and even biological observations through equations and formulae. For instance, such formulae helped engineers commit to the outcome even before a building is built or a pipeline is extended. Over time, many equations were made redundant as fewer formulae became applied more often to explain a larger number of observations. Stephen Hawking, in his book, states that he expects the time will come soon when it will take only one formula to explain everything. It will be the mother formula. It will be so simple that we would be amazed that it was never thought of earlier. Could the answer be residing in the revelation process? Prophets had revelations through which the words of God were able to pass to people. Scientists, through revelations, were able to come up with theories and make remarkable inventions. All humans from time to time are able to integrate existing components or things in a way never done before and thereby create innovations. Neither age, nor skill or experience makes a difference. Wolfgang Amadeus Mozart did not acquire skill, nor experience to compose his first music work at the age of five.

It is the widely-perpetuated myth that humans only use less than 10% of their brains (Radford, 2007). By association, it is often suggested that by some process a human being may harness this unused potential and in so doing inherit several magnitudes of more intelligence. Though many factors of intelligence may be increased with training, the idea that large parts of the brain remain unused is without a substantial foundation. Although there are still many mysteries regarding brain function, every part of the brain has a known function. The pineal gland acts as the third eye or portal to download information that is stored in particles energy state around us in the universe. Photons particles for instance follow quantum behavior and, as such, could exist in two locations at same time. If this were to occur, would the information flow be uni- or bidirectional? It would be a waste of capability if the information were to flow in one direction only.

How many times have you been chatting with friends and colleagues, when suddenly the person sitting next to you speaks the words that you were about to say? There are people with a strong ability that we used to call the 'sixth sense'. It enables them, unintentionally, to learn of others' thoughts. The question became; what would the rest of the brain be used for? Could it be that a part of the brain is used to i) set the address of the universal information to be fetched and/ or stored and ii) resolve the header information so that the right code page is applied and the type of data; audio, visual, textual,... etc, is defined? When the decoding key is not the correct one, received information becomes meaningless or a mere hallucination, as in some of our dreams. When the decoding key is the correct one, the downloaded information becomes meaningful and results in inventions, innovations, or time-traveling visions. In most holy books we read about destiny as being predetermined and prewritten or pre-stored in the universe around us.

The human organs behave more 'classically' and less 'quantum-physically'. Unseen and undetected, the human soul could, on the other hand, behave more 'quantum- physically' and less 'classically'. Could the human soul be resident in the part of the brain that we know little about? Following wave-particle duality, could the human soul be located at that ambiguous, spiritual part of the brain and at the same time at a location in the universe where the information to be exchanged is residing? The pineal gland would then be responsible to confirm the connection and act as a gateway to upload or download such information. The brain's process of revelation, invention or innovation thus becomes clearer, and the 'Theory of Everything' becomes closer to reach. The question is no longer 'What is the formula that one applies to fix a problem'? But, rather, 'Where, in the time-space universe, is the location in which the information on a specific fix is stored, and which code page applies'? To the naked eye, a sequence of qubits could appear as *'simple data'* or a set of *'program instructions'*. Therefore, the downloaded information may be simple data or sequence of instructions for the logical part of the brain to unfold. Gravity sucks energy out of the quantum fields. Could such programs excite the energy of the quantum field, which is always almost positive? If so, a balancing negative energy of gravitational attraction will result. A mix of energy and gravity in a neatly set of program instructions could very well perform telekinesis, remote healing or containment of viral or microbial contamination.

Revelation experience comes to most of us. As noted earlier, we sometimes refer to it as the 'sixth sense'. Since the sixth sense is a very weak sense, many people go through life without realizing their true potential capabilities. Chatting with a few acquaintances and friends in the summer of 2008 in Essen, Germany about areas of economic development in Somalia, we discussed how the country was devastated by wars and disputes for the past two decades. A senior diplomat of the Punt land State of Somalia, Mr. Saeed Fareh, spoke of a Salt mine that was deserted since the beginning of World War II. The mine is located near a Somali village called 'Eil'. Mr. Fareh tried to emphasize the name of the place, so he said that Eil sounds the same as the trailing syllable of Ishmael, Gabriel, Michael, and many names of prophets and angels who are mentioned in the holy books. It suddenly hit me that a location on the map close to Eil could be the place where Noah had lived. "That is the place where he built the Ark" I said. After a moment of silence, the diplomat's face turned crimson red and his eyes opened up in surprise as he asked what made me say so. It occurred to me that, given a sudden tilt of Earth's magnetic core, as I will explain later in the second and third chapters of this book, the spin speed of the planet would have jumped threefold resulting in sudden rise in ocean level at the Equator and the Tropics regions which would have created a giant ocean wave of epic proportion. A tsunami would have hit the horn of Africa, and would have lifted Noah's Ark, which was docked on dry ground. The sudden tilt and drop in the strength of the Magnetic Field would have caused a sudden melting of the ice caps that resulted in increased sea level, which would have soon transformed into vapour thanks to the thermal energy reaching the surface of the Earth. The vapour would have gathered into clouds and rained on Earth for days and weeks, specifically for forty days, as mentioned in the Ark of Noah story in the holy books. I tried to reason my revelation at the meeting but failed to answer why I pointed my finger to that specific spot on the map!

The Somali diplomat immediately explained that at the specific spot where I pointed my finger on the map; there exists a wide passage where there are the "Most currently known animal and bird footprints imprinted since millennia". As a nearby volcano spread ashes into the air, the footprints became fully covered with ashes that solidified their exact shapes and preserved the print for thousands of years. The holy books tell us that pairs of animals and birds marched onto the Ark of Noah. This could very well have taken place some 10,500 years ago. Thanks to the cold climates of Europe and Asia, it would have been only possible to have most of the animals of our planet living in Africa's moderate climate. It could, therefore, be that the Ark was built and took off from somewhere in Africa, and not Europe nor Asia.

There are discoveries that the Ark of Noah landed in Turkey over the mountain of Ararat. The Red Sea must have been connected to the Mediterranean through other means than the 150-year-old Suez Canal, or perhaps water level rose high enough so that the two seas were temporarily connected. History tells us that several canals were dug between 610 BC and 767 AD in an effort to connect the Nile to the Red Sea and neighboring Lakes (Egypt, 2006). Ancient Egyptians had traveled between the two seas through a canal that connected the Suez Gulf to the Nile. Initially it was called the Sesostris (Senusert) Canal. Through the Nile, ships could have reached the Mediterranean from the Red Sea. Higher water levels due to continuous and heavy rain could also have made the navigation self run. It is therefore obvious that climate change has shaped the evolution of modern mankind over 70 thousand years. It has expanded the power of its brains to open up to everlasting wealth of information and knowledge and populate the planet. Are we on the path of another cycle of climate change that will lead to further development of mankind abilities? What are the observations, causes and consequences? What is the time line? And whether we are prepared?

As we reflect on it, we find that each discovery was initially based on a thought or idea that inspired the mind of the discoverer to try to verify and validate, through several means such as developing an analytical and/ or a philosophical model, as well as pursuing physical evidences and associate proven theories. As I do not have the adequate resources to establish an analytical model for planet Earth, I will try to establish a philosophical model by mosaic-ing many pieces together from various disciplines (electricity, magnetism, history, geology, etc) to explain how the planet functions and how it is influenced by the surrounding space. I will also try in the following chapter to leverage established scientific theories to qualify each component of the philosophical model of planet Earth. I remain indebted, at all times, to the pineal gland or the third eye for guiding me to reach what you are about to read in the coming chapters.

The Earth

'Irreversible climate change due to carbon dioxide emissions' is a paper written by (Solomon, Plattner, Knutti, & Friedlingstein, 2009). It shows that the current climate change taking place due to increased carbon dioxide concentration is largely irreversible for 1,000 years after emissions stop. But, climate change does not justify the alarming increase in the number of Earthquakes all over the planet in recent years. As the frequency of Earthquakes, tectonic disturbances and oceanic volcanoes is plotted over the past forty years, we find that the annual frequency of Earthquakes, above 6 Richter, has risen more than double starting 2007. Some scientists are blaming the reversal of the magnetic poles. However, the question remains; why are the magnetic poles reversing? And, could the weakening of the magnetic field results in climate change and be also related to the alarming increase in frequency of Earthquakes?

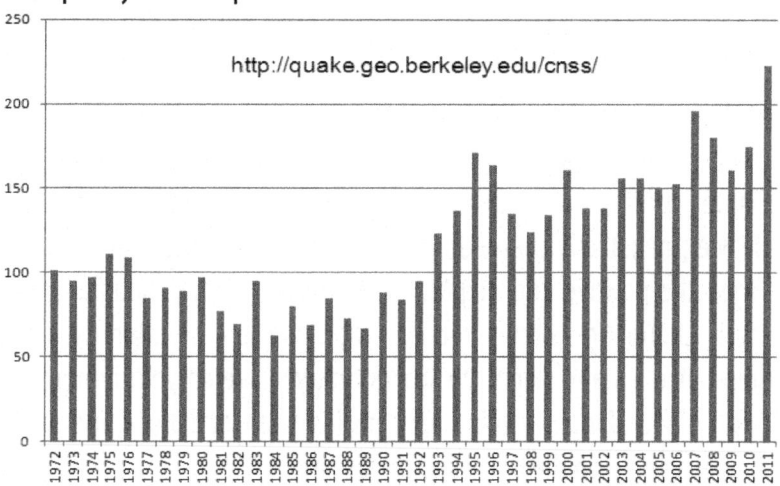

Figure 6- Number of Earthquakes higher than 6 Richter since 1972

Many are predicting that a cataclysmic event has either been avoided or is being initiated. The same is happening in Western and Northwestern America, where more than 600 small Earthquakes have been registered in 2005. According to scientists, hot spots such as Sumatra, Yellowstone, and some others are showing signs of extreme disturbance and harmonic tremors. It is now confirmed that we have a higher frequency of occurrence of powerful hurricanes and floods. Could all of these incidents be linked through a single chain of cause and effect? And, what could that be?

Strange sky sounds. Have you ever got a static electricity shock? Or seen sparks when you take off your jumper? Well, lightning follows the same principle, but on a much bigger scale. Lightning is an electric current. To make this electric current, first you need a cloud. When the ground is hot, it heats the air above it. This warm air rises. As the air rises, water vapour cools and forms a cloud. When air continues to rise, the cloud gets bigger and bigger. In the tops of the clouds, temperature is below freezing and the water vapour turns into ice. The cloud then becomes a thundercloud. Lots of small bits of ice bump into each other as they move around. All these collisions cause a buildup of electrical charge. Eventually, the whole cloud fills up with electrical charges. Lighter, positively charged particles form at the top of the cloud. Heavier, negatively charged particles sink to the bottom of the cloud. When the positive and negative charges grow large enough, a giant spark - lightning - occurs between the two charges within the cloud. This is like a static electricity sparks that you see on your jumper, but much bigger. Most lightning happens inside a cloud, but sometimes it happens between the cloud and the ground.

Charged particles arriving from the sun, such as protons are normally trapped by the Magnetic Field that fills the Thermosphere layer some

100 to 800 km (62 to 500 miles) above sea level. Protons do not remain still but spiral along the Magnetic Field Force lines moving between the two Magnetic Poles. The weakening of the Magnetic Field by 10% over the last 150 years and by 5% over the last 30 years alone indicates an accelerated downfall of the Magnetic Field. While some protons get trapped in the Thermosphere layer, others find it easier to penetrate the weaker Magnetic Field to reach the Troposphere layer where weather, major wind systems, and cloud formations mostly occur and extend from sea level up to 17 km (10 miles) above sea level.

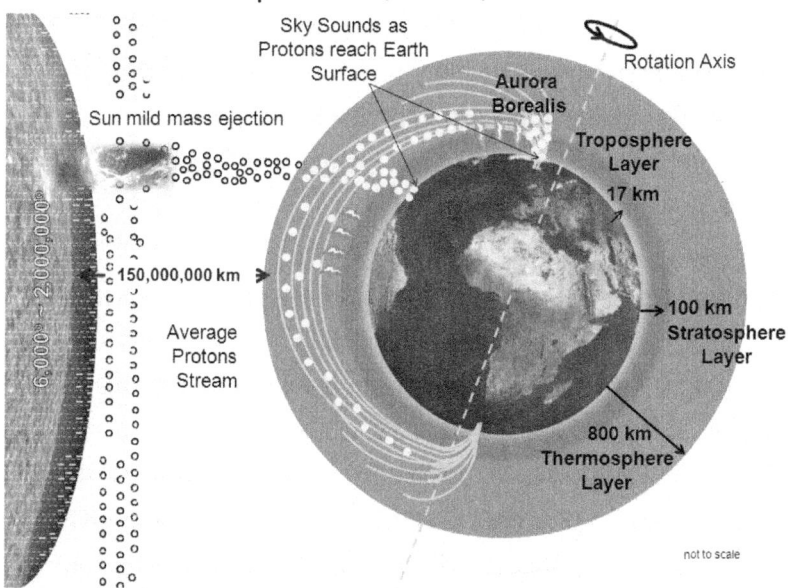

Figure 7- Arrival of sudden high levels of Protons

The Sun ejects from time to time big chunks of mass and energy, that we call Coronal Mass Ejection (CME). Some of these sudden CMEs could fire their way in the direction of Earth at random. The penetration of sudden swamps of positively charged, high-speed, protons into the negatively charged, low-speed electrons filling the bottom of clouds generate micro bolts of lightning on sub-atomic level. The grumbles and growls we hear as strange sounds and roar from the sky do actually come from the rapid expansion of the air surrounding the micro lightning bolts. It is commonly known that protons spiraling along the Magnetic Field Force lines, moving between the two magnetic poles, will have to come closer to Earth's surface as they enter into the Troposphere layer at the Magnetic Poles sites, thus creating the Aurora Borealis lights. It was researched most recently that they also generate sky sounds at close proximity to the Earth's surface[3]. In a nutshell, the protons penetrating the Troposphere, whether at the Magnetic Poles or into the Magnetic Field Force lines, at a region spot that happen to fall in the path of a Sun mild CMEs produce sky sounds as observed lately in many parts of the world. It is therefore that occurrences of Earthquakes and preceding strange Sounds from the skies, as reported in December 2004 before the Asian Earthquake/ Tsunami, as well as in other recent incidents, are not related.

Sudden death of birds and fish. Just before midnight on new year's eve of 2011, anywhere from 1,000 to 5,000 red-winged blackbirds and starlings fell from the sky within a one-mile area over the town of Beebe, Arkansas[4]. Recently, a mass fish kill, in which an estimated 100,000 drum fish washed up on a twenty-mile stretch near the town of Ozark, Arkansas, which is about 200 km (125 miles) away from Beebe. And then, around 500 red-winged blackbirds, starlings, and grackles fell to their deaths over a quarter-mile stretch of highway near Labarre, Louisiana, which is 576 km (360 miles) from Beebe and 720 km (450 miles) from Ozark. At the same time, hundreds of what were most likely jackdaws fell to the ground all over Falköping, Sweden! The increase of protons reaching the surface of the Earth results in

---

[3] http://www.aalto.fi/en/current/news/view/2012-07-09/
[4] http://io9.com/5725175/why-are-thousands-of-dead-birds-suddenly-falling-from-the-sky

increasing cases of depression and cancer risks, damage to central nervous system and cataracts as well as other diseases. The acute effect of a whole-body proton radiation on the function of leukocyte populations in the spleen and blood were examined in small mass animals. Adult female mice were exposed to a single dose (3 Gy, where Gy is Gray or unit of Radiation) of protons and were killed humanely at six different times thereafter. A typical schedule of treatment of human cancer via radiation for adults, for comparison, is 1.8 to 2 Gy per day, five days a week. I tend to believe that the events of sudden death of Birds and Fish in the recent years could very well be attributed to the damage of their central nervous system due to being exposed to high doses of protons radiation at the time of sudden CMEs mass and energy reaching Earth.

Aluminum versus carbon composites. Charged particles arriving from the Sun are trapped along the magnetic field force lines. The continuous weakening of Earth Magnetic field allows a percentage of such charged particles, namely protons, not to get trapped and to penetrate into the Troposphere layer where most air travel takes place. As a result, increased proton level at the surface of Earth has been detected in recent years. Additional doses of protons are also detected and are mostly associated with sudden CMEs that randomly fire from the Sun towards of Earth. The rush of such additional protons, at small instants, leads to increased penetration into the magnetic field and arrival to the Troposphere of instant surges of positively charged electric particles. Jet airliners used to employ Aluminum as the main material in their primary structure including the fuselage. Carbon composites material is introduced lately at 50% rate into the composition of the structure of jet airliners. It is well-known that Aluminum enjoys 1,000 times better conductivity than Carbon composites. The decrease of Aluminum and increase of Carbon composites into the structure of jet airliners leads the fuselage to become less conductive and less able to electrically ground the arrival of random surges of charged particles. Fuselage poor conductivity allows charged protons to penetrate into the main cabin. This could cause increased radiation doses as well as the possibility to damaging electronic circuits and electric batteries; causing fire! And, as suggested in Chapter 4, a conductive enclosure is the best protection for such electric circuits.

Climate and pole changes. There are evidences that the climate is continuing to change, both in increasing global average temperature resulting in what is called Global Warming. At the same time the climate is exchanging its patterns. North America is getting warmer and shorter winters while Siberia, Europe and the Middle East are getting colder and longer ones. The records collected and historical information going back 300 years show a consistent temperature model. However, we should not focus on the syndrome and fail to notice the root that is causing such changes. Watching a documentary, back in 2002, on the BBC channel, it was reported that the magnetic pole has shifted some 104 km (65 miles) away from its previous location close to the North geographic pole in just a few years. Many scientists have stated that the Earth's magnetic pole in the North is drifting away from the Arctic Ocean, so quickly that it could end up in Siberia sometime in the very near future. The shift could mean that Alaska will lose its northern lights, or Auroras, which might then be more visible in areas of Siberia. The magnetic poles are distant from geographic poles that mark the surface points of the axis of Earth's rotation about itself. An imaginary line joining the magnetic poles is currently inclined by approximately 10° from the planet's axis of spinning. Currently, the south magnetic pole is located near the Earth's geographic North Pole and the north magnetic pole is situated near the Earth's geographic South Pole. The two magnetic poles wander independently from one another and are not at directly opposite positions on the planet (Canada-Natural-Resources, 2005) (Australian-Antarctic-Division, 2002). Magnetic poles are known to migrate and occasionally swap places. "This may be part of a normal oscillation. It will eventually migrate back toward Canada", told Joseph Stoner, a paleomagnetist at Oregon State University, at a meeting of the American Geophysical Union in 2005 in San Francisco.

Navigation at sea and on land was highly dependent on the magnetic field until recently. Even though, today, we have the pinpoint accuracy of the satellite-based Global Positioning System (GPS), many still rely on their compass needles pointing to the magnetic pole in the North. The position of the magnetic pole changes and there is evidence that it is changing at an increasing rate. It was Larry Newitt's job at the Canadian government-funded Geolab to track the wandering magnetic pole[5]. Every few years he undertook a seven-hour flight from his base in Ottawa to Resolute Bay, the closest inhabited spot to the magnetic pole, then a three-and-a-half-hour flight north in a Twin-Otter aircraft. "Today, the magnetic pole in the North is at sea and the expedition can only be done at the end of the winter when the sea is frozen". Placing magnetic sensors on the ice, the expedition attempts to surround the magnetic pole and triangulate its correct position. Each year they go back and find that it has moved. "We're following it across the ice", Larry Newitt told BBC News Online. "It jumps around from day to day and year to year and we have to keep track of it". Measurements in

---

[5] http://news.bbc.co.uk/2/hi/science/nature/2889127.stm Is the Earth preparing to flip? By Dr David Whitehouse (2003).

1904 of the magnetic pole's position up in the north, by explorer Ronald Amundsen, put it in roughly at the same place as an earlier, though less accurate, measurement made in 1831 by the British explorer John Ross. Since then it wandered slowly northward until about thirty years ago, when it started behaving differently. "There was a slow drift northward, but then it started to move faster. It is now moving northward, away from Canada to Siberia, at a rate some four times faster than it used to", said Newitt. Soon, he added, "expeditions to the magnetic pole would become more difficult as it moved out of range of the Twin-Otter aircraft". However, something captured my attention when I looked at Figure 8; the total area of the ice cap was greater in 1979 than that in 2003. However, the encircled area, which represents a bordering region to north of Siberia had larger ice cap in 2003 than in 1979! It seems that the ice cap is melting at the Canadian side where the magnetic pole is moving away from, and is building up at the Siberian side where the magnetic pole is moving towards, Therefore instead of crying out that

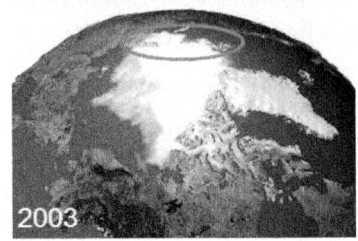

Figure 8- Ice Cap Comparison over a period of 24 years

the ice cap is melting, shouldn't we better state that the ice cap is melting but is also relocating!

Wandering poles are confirmed by studies which show that the strength of the Earth's magnetic shield has decreased 10 percent over the past 150 years. During the same period, the south magnetic pole, up in the north, has wandered about 1,100 km (685 miles) into the

Arctic Circle[6]. "The rate of the magnetic pole's movement has increased in the last century compared with fairly steady movement in the previous four centuries", said Joseph Stoner and the Oregon researchers. The Oregon team examined the sediment records from several Arctic lakes. Looking at the sediment records of the Earth's magnetic field at the time; scientists used carbon dating to track changes in the magnetic field. They found that the south magnetic pole had shifted significantly in the past thousand years. It generally migrated between northern Canada and Siberia, and has occasionally moved in other directions. The rate of shift of the magnetic pole is on the increase; moving from northern Canada towards Siberia. The south magnetic pole in the geographic north was first discovered in 1831, and when it was revisited in 1904, explorers found that it had moved by 50 km (31 miles). For centuries, navigators using compasses had to learn to deal with the difference between magnetic and geographic poles. A compass needle points to the magnetic pole, not the geographic pole. But recently, the rate of movement has increased to become few hundreds of miles in just less than a decade. Something

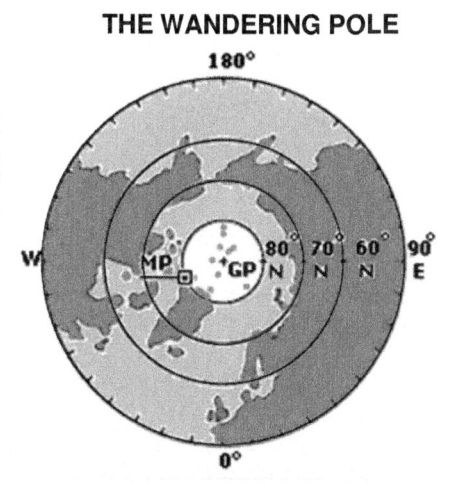

THE WANDERING POLE

+ **GP** Geographic Pole
• **MP** Current magnetic pole
• Past magnetic pole positions

Figure 9- Rock samples have determined the changing position of the magnetic pole over the past 7,000 years

---

[6] http://news.bbc.co.uk/2/hi/science/nature/4520982.stm; "Magnetic north pole drifting fast," BBC (2005)

unexplained is occurring to the Earth's magnetic field. Could there be a relation between the shift of the magnetic pole and the shift of the Ice Cap; both are moving towards Siberia? And why the ice cap in the Arctic is melting while in the eastern Antarctic is growing[7]?

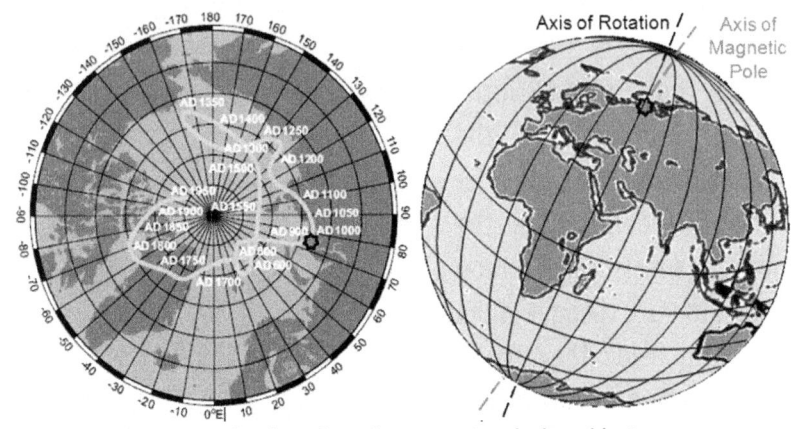

Figure 10- The closer from the magnetic pole the colder it gets

I found an intriguing answer by tracing the magnetic pole ancient movements together with documented climate conditions at ancient times. It is apparent that the closer we come to the magnetic pole, the colder it gets. Arab ancient texts speak of an abrupt climate change in ancient times[8]. The manuscripts come from ancient texts that describe, in detail, a bitter cold wave that occurred between the years 900 AD and 950 AD, in Arabia. Research has been brought to light by a team of scientists from the University of Extremadura. The article published describes, how in 900 AD, the territory where it is now Iraq and Syria suffered periods characterized by a very unusual cold, with temperature below zero. According to Fernando Domínguez-Castro, who is carrying research at the University of Extremadura in Spain, the study has brought to light key data that better interpret the current

---

[7] http://www.csmonitor.com/Environment/2008/0110/p14s01-sten.html
[8] http://www.livescience.com/18650-weird-weather-ancient-baghdad.html

climate change. "We were fortunate to find ancient sources of direct data that are traditionally neglected by many climate scientists" added Dominguez-Castro[9]. Some researchers think this may presage a magnetic reversal, in which the south and north magnetic poles flip. However, this does not justify why the strength of the magnetic field is weakening. Scientists admit that there are things going on, deep beneath our feet, which they do not understand. I add, that there are also celestial occurrences going on, above our heads, which we need to be made aware of. Both types of changes could be inter-related and have profound consequences over life on Earth as it will be manifested over the next few lines.

Pole reversal is a consequence of the magnetic core completing a half-circle tilt. Studies confirm that the magnetic field is getting weaker. David Kerridge, of the British Geological Survey, told BBC News Online, "There is strong evidence that the field is decreasing by about 5 percent per century". We studied at school that it takes a magnet to pull, repel or tilt another magnet. Therefore, we should expect a magnetic core of some planet to be influencing our own and causing it to tilt, of course; assuming we have a permanent one! Since the incident appears to be unprecedented in the recorded history, such a planet cannot be one of the planets currently known to orbit our solar system. One hypothesis could be that the magnetic force lines of the core of the Earth are fluxing deeper into space. They are trying to reach out and bond with the magnetic force lines of such an intruder that is approaching the solar system. Due to the magnetic pull of such a planet, the Earth's magnetic force lines could be further bulged and protruded while travelling around Earth between its two magnetic poles. This makes the magnetic field to appear weaker when measured

---

[9] Added in the 2012 print edition of this book

from the surface of the Earth, except when measured at the exact location of north magnetic pole in the Antarctic, where the strength of the flux-out magnetic force lines remains unchanged.

Some researchers suggest that it could be the start of a magnetic reversal, and that it will take a few thousand of years to have the north and south magnetic poles reversed. I believe that we will be able to determine the end position and the travel time of the magnetic poles only when the reason for the tilting is detected and/or historical recurrence is confirmed. When the rate of movement of the magnetic pole, over the past four years has quadrupled, it is not unlikely that the tilting rate will continue to grow even faster. Some scientists said that it may not be a few thousand years before the poles are reversed. What if, it is only a few years, or even a few months or days? There is evidence, in petrified woods in Illinois and Kentucky that giant club moss trees were stressed due to a sudden climate change and vanished overnight. They were replaced by weedy fern vegetation. "On average, pole reversal occur about every 250,000 years", said Dr. David Whitehouse, BBC News Online Science Editor. "However, it has been 750,000 years since the last reversal. Are we overdue? Or are the geological records incomplete?" he added. But, not until we understand the **relation between the change in climate and the tilting and weakening of the magnetic force lines,** that I can accept the statement of Dr Whitehouse!

Why Earth temperature varies at sea level? Earth has an axial tilt of about 23.44° (23° 26'). The axis is tilted in the same direction throughout the year. However, as the Earth orbits the Sun, the hemisphere that is tilted away from the Sun will gradually become tilted towards the Sun while moving on a near circular orbit, and vice versa for the other hemisphere. This effect is the main cause of the four seasons. The hemisphere, facing the Sun, experiences more hours of sunlight each day.

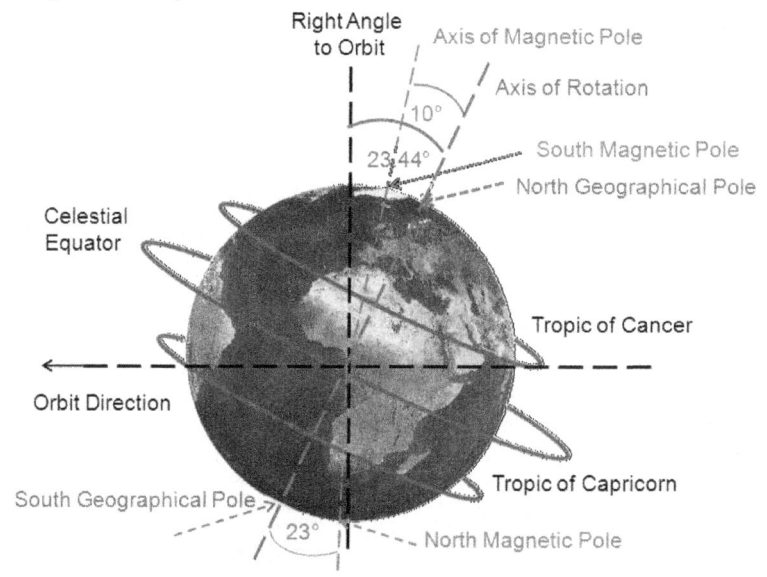

Figure 11- Axial tilt of the Earth

The Tropic of Capricorn, or Southern tropic, is one of the five major belts or circles of latitude. It marks a region of homogenous temperature on the map of the Earth. It lies 23° 26' south of the Equator, and marks the most southerly latitude at which the Sun appears directly perpendicular on December 21 in an event that is called the Winter Solstice. Due to Earth slight wobbling around its axis, the Winter Solstice is very slowly moving away from December 21. Equally, in the Northern Hemisphere, equivalent of the Tropic of Capricorn, there is the Tropic of Cancer at which the Sun appears directly perpendicular on June 21 in an event that is called Summer Solstice. The region north of the Tropic of Capricorn and south of the Tropic of Cancer is known as the Tropics. Therefore, it is the case that the Sun's perpendicular appearance on the surface of the planet is confined and is in constant forward and backward movement between the two tropics. Some believe that the Equator experiences the highest temperature since it is thought to be closer to the Sun than any other region of our planet! But, the Earth axial tilt does not make the Equatorial region any closer to the Sun than the southern Middle Latitude region on a Winter Solstice day for example.

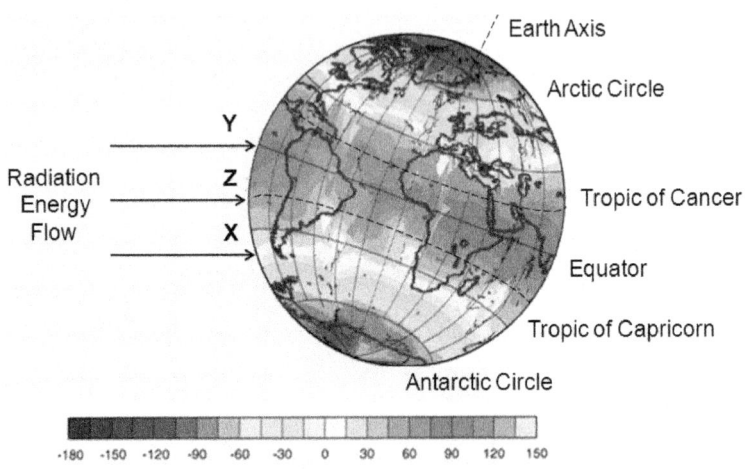

Figure 12- The level of thermal energy that reaches the surface of the Earth is independent on the distance between Earth's regional surface and the Sun[10]

The closest to the Sun, on Figure 12, would be the region tangent of the orbit plateau of the solar system i.e. on the Tropic of Capricorn. Some attribute the angle of projection between the solar rays and the surface of Earth to influence the temperature variation! The angle of projection starts from 90° at the region tangent to the orbit plateau of the solar system and grows smaller until it reaches 0° at the region perpendicular to the orbit plateau of the solar system. This does not explain why the surface region that is located at 90° (degree) with respect to the solar energy, i.e. at the Tropic of Capricorn during a Winter Solstice day, has less temperature than the Equator's, where the

---

[10] Temperature Graphic by D. Hartmann and M. Michelsen, University of Washington

temperature is currently highest while the angle of projection is less than 90°! Lying between the two tropics, and if the angle of projection and/ or proximity to the solar energy are the drivers of high temperature, the Equator could only have the chance, once every six months, to be situated at a right angle, distance-closest to the Sun. This should not make of it the hottest place on Earth all year round. So, why it is?

As illustrated on Figure 12, on a Winter Solstice day, when region **y** (Equator) is at the same distance and angle of projection from the Sun as region **x** (Middle Latitude); why then, do these regions have a difference in temperature? And what makes region **y** (Equator) the hottest place on Earth all year round; while the nearest region to the Sun, region **z** (Tropic of Capricorn), experiences a lower temperature? **You may also start to wonder, what makes temperature differences over the surface of the Earth when measured at sea level!**

We understand that the magnetic field force lines converge at the magnetic poles and diverge at mid distance between the two magnetic poles or what I call the magnetic equator. The charged particles arriving from the sun find it hard to penetrate the magnetic field and tend to spiral along magnetic field force lines in go and fro mode between its two magnetic poles. The distance travelled and the speed of such protons reach maximum at the magnetic equator and reach minimum at the magnetic poles. During such travel, protons collide with air molecules of the Thermosphere layer that is situated 100-800 km (62-500 miles) above the surface of the Earth. The collisions thermal energy is proportionate to the helical distance travelled as well as the speed of protons; it reaches 2,000° Celsius above the magnetic equator and 500° Celsius above the magnetic poles. We understand that the Sun is situated 150 million km (93 million miles) away from

Earth and that it has a surface temperature of 6,000° Kelvin. Most Scientists claim that the thermal radiation arriving from the Sun is the source of the energy that heats the Earth surface. But, what about the thermal radiation that reaches the surface of Earth from the Thermosphere layer? Let us see; only a fraction of the total power emitted by the Sun falls on an object in space, the Earth, which stands at a distance from the Sun. The solar irradiance in Watt/m2 is the power density incident on Earth due to radiation from the Sun. At the Sun's surface, the power density is that of a blackbody; a body that emits radiation energy uniformly in all directions per unit area normal to direction of emission, at about 6,000° Kelvin. The total power from the Sun is this value multiplied by the Sun's surface area. However, at some distance from the Sun, the total power from the Sun is then spread out over a much larger surface area and therefore the solar irradiance on an object in space decreases as the object moves further away from the Sun. For instance the total power from the Sun reaching Mars at 227 million km is much less than that reaching the Earth at only 150 million km away from the Sun (Earth Orbit Radius= D).

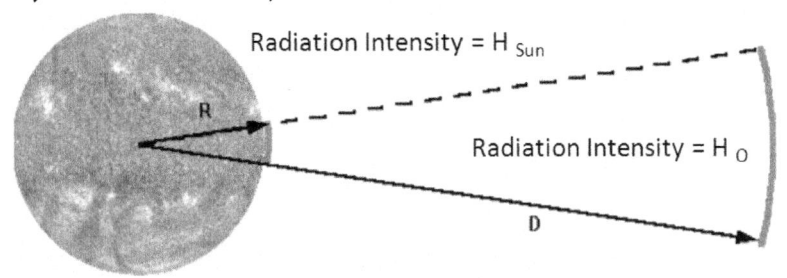

Figure 13 - Radiation Intensity due to the Sun

The solar irradiance on Earth at 150 million km D from the Sun is found by dividing the total power emitted from the Sun by the surface area over which the Sunlight falls. The total solar radiation emitted by the Sun is given by σT^4, as defined by the Boltzmann's blackbody equation multiplied by the surface area of the Sun ($4\pi R^2_{Sun}$) where R_{Sun} is the radius of the Sun. The surface area over which the power from the Sun falls will be $4\pi D^2$. Where; D is the distance between the object and the Sun. Therefore, the solar radiation intensity, H_{E-S} in (Watt/m2), incident on the Earth looks as follows:

$$H_{E-S} = \frac{R^2_{Sun}}{D^2} H_{Sun}$$

where;
- H_{E-S} is the radiation intensity (in W/m^2) at the Earth's Troposphere due to radiation received from the Sun.
- H_{Sun} is the radiation density at the Sun's surface (in W/m^2) as determined by Stefan-Boltzmann's blackbody equation $E = \sigma T^4$; where $\sigma = 5.67 \times 10^{-8}$ W/m^2 x K^4
- T is the temperature of the surface of the Sun at 6,000° Kelvin.
- R_{Sun} is the radius of the Sun in meters as shown in Figure 13; and
- D is the distance from the Sun to the Earth's surface in meters as shown in Figure 13.

It is therefore found that the radiation intensity reaching the Earth from the Sun is 1,366 Watt/m^2.

It is measured that the Thermosphere temperature varies between 500° to 2,000° Celsius depending on the Sun's activity and the strength of the magnetic field force; where it is strongest, at the magnetic poles, the Temperature is 500° Celsius, and where it is weakest at the mid region between the magnetic Poles (i.e. the magnetic equator) it reaches 2,000° Celsius.. Following a similar model to that of the Sun/Earth radiation as explained above, let us build a model Thermosphere/Earth radiation, the thermal radiation or heat exchanged at the surface of the Earth from the Thermosphere would be;

$$H_{E-T} = \frac{\sigma T^4 \times S_{\text{heat ellipsoid in thermosphere}}}{S_{\text{heat ellipsoid reaching Earth surface}}}$$

Where,

- H_{E-T} is the radiation intensity (in W/m^2) at the Earth's Troposphere due to radiation received from the Thermosphere.
- T is the temperature of the mid distance between the two magnetic poles at the Thermosphere and is taken at an average of 1,800^0 Kelvin.
- $S_{\text{heat ellipsoid in thermosphere}}$ is the highest thermal radiation region of the Thermosphere that is modeled as an ellipsoid of radii equal to the weakest magnetic contour at 24 mTesla (550 km, 600 km) and height of 10 km (where most of the Sun's charged protons get trapped), and is calculated as follows, $S = 4\pi[(a^p b^p + a^p c^p + b^p c^p)/3]^{1/p}$; where p=1.6075 and a= 550 km, b= 600 km, c= 10 km.
- $S_{\text{heat ellipsoid reaching Earth surface}}$ is the reach of the highest thermal radiation area of the Thermosphere to the surface of the Earth that is modeled as an ellipsoid of radii of 9,000 km (i.e. 1/4 Earth circumference) and height of 360 km from Earth's surface, and is calculated as per above surface formula S, where p=1.6075 and a= 9,000 km, b= 9,000 km, c= 360 km.
- The selection of the highest thermal radiation as an ellipsoid within the Thermosphere layer is driven by the shape of the temperature map measured as shown in Figure 14 for the Thermosphere [11].

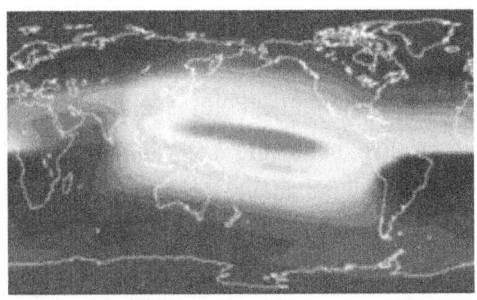

Figure 14 – Thermosphere Temperature Map

---

[11] http://ccmc.gsfc.nasa.gov/models/modelinfo.php?model=CTIPe

It is therefore found that the radiation intensity that reaches the Earth from the Thermosphere is 2,412 Watt/m^2. This means that the thermal radiation reaching the Earth's surface from the Thermosphere is approximately 1.8 fold stronger than that reaching the Earth from the Sun directly. The trapped, oscillating Sun's protons between the two magnetic poles, day and night, keeps the Thermosphere thermal radiation uninterrupted, though decaying. Such a phenomenon keeps the Earth surface temperature safe from sharp drop at night.

As the Sun ejects mass-energy of heavy particles such as electrons and protons, the magnetic field at the Thermosphere layer shields such energetic bodies. The trapped protons, full of kinetic energy, have no place to go but to spiral along the magnetic field lines while they move between the two magnetic poles.

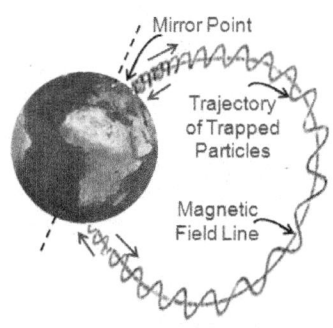

Figure 15- Trapped Protons

As protons encounter regions of stronger magnetic field where field lines converge, their spiral-radius is shortened and their speed is slowed down. The protons could reverse paths at the magnetic poles. This could cause the protons to bounce back and forth [12] between the two magnetic poles. It also keeps the thermal radiation, coming from the Thermosphere, to gradually decay above the sky of a specific region of Earth's surface for the rest of the day including after sunset. As protons spiral around the magnetic field force lines, they reach the maximum spiral-radius and speed at mid region between the two magnetic poles where the magnetic field intensity is lowest. The protons reach the minimum spiral-radius and speed at each of the magnetic poles, where the magnetic field intensity is highest.

---

[12] http://www-spof.gsfc.nasa.gov/Education/Iradbelt.html

Collisions between such spiraling protons and the Thermosphere air molecules at various speeds produce thermal energy and temperatures that are proportionate to the protons speed and spiral-radius motion. Temperature is found to reach 500° Celsius above the magnetic poles and to gradually escalate to reach 2,000° Celsius above the magnetic equator.

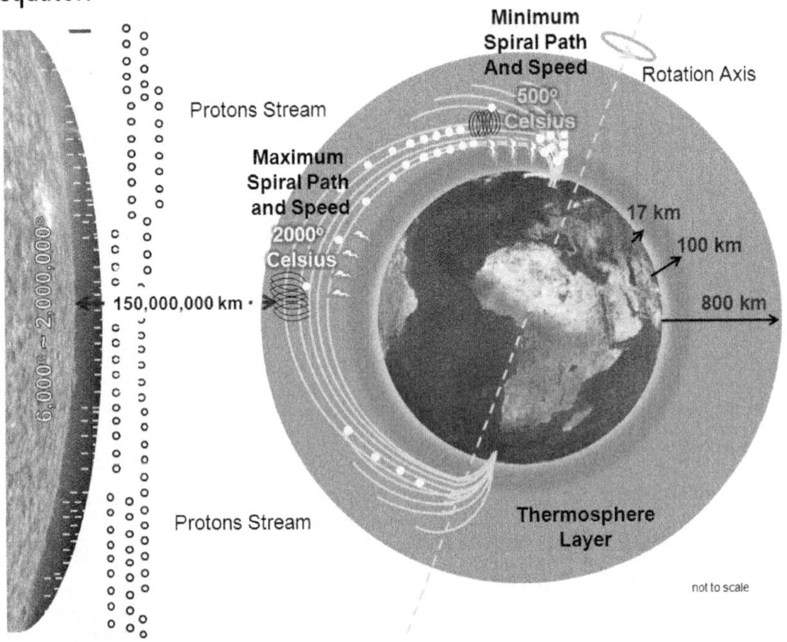

Figure 16- Thermal energy arriving from the thermosphere

This makes the region of magnetic equator to always maintain the highest temperature on the surface of the planet and for the regions of the magnetic poles to maintain the lowest temperature. The weakening of Earth's magnetic field leads to an increase in the Protons speed and spiral-radius motion around the magnetic force lines. More chances are created for the Protons to collide with Thermosphere air molecules at a higher speed. The impact of stronger collisions results in higher thermal energy reaching Earth's surface; **Global Warming** is thus observed. It is imminent, therefore, that a change in the temperature pattern and precipitation map of the planet will follow any change or repositioning of the magnetic poles and the associated magnetic field strengths. Several questions remain to be answered. Shall the magnetic poles reversal be carried out gradually or abruptly? Shall it commence a gradual tilt only to escalate as time goes by? Will the magnetic pole reversal, if at all, be carried out on distinct stages separated by pause intervals? Since the temperature belts are imaginary lines running perpendicular to the line connecting the two poles and enjoy homogeneous climate each; will the temperature belts also tilt along with the tilting of the magnetic poles?

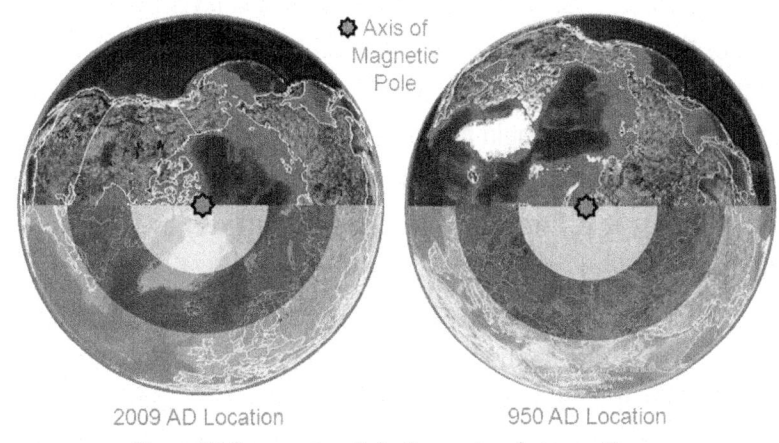

Figure 17- Temperature Belts Comparison between Times

If we reflect on **Rate of Shift** section discussed earlier in this chapter, and circle the Temperature Belts around the magnetic pole, we shall find an obvious climate zone similarity between current date Europe and ancient date northern Arabia in the years between 900 AD and 950 AD as shown in Figure 17. This is an obvious indication that Temperature Belts follow the Thermosphere temperature which follows the intensity of the magnetic field as well as the locations of the two magnetic poles.

The Earth's structure is very close to an oblate spheroid (Milbert & Smith, 2007); a rounded shape with a bulge around the Equator. The average diameter of the reference spheroid is about 12,742 km (7,963 miles). The circumference measures 40,000 km (25,000 miles). The meter was originally defined as 1/10,000,000 of the distance from the Equator to the North Pole through Paris, France. The rotation of the Earth creates the equatorial bulge so that the equatorial major axis diameter is 43 km (27 miles) larger than the pole-to-pole minor axis diameter. The largest local deviations in the rocky surface of the Earth are Mount Everest, at 8,848 m (29,028 ft) above sea level and the Mariana Trench, 10,911 m (35,798 ft) below sea level. Because of the bulge, the feature farthest from the center of the Earth is actually Mount Chimborazo in Ecuador. Its summit is 6,268 m (20,565 ft) above sea level, which we should not forget is also bulged along the Equator. **The bulge of the planet's Crust and Oceans sparked a question on whether it is limited to the Earth's surface, or if there is a bulge at every geological layer beneath?** Gravity tends to contract a celestial body into a perfect sphere, the shape for which all the mass is as close as possible to the center of gravity. Rotation causes a distortion from this spherical shape. A common measure of the distortion is the flattening, which can depend on a variety of factors including the size, angular velocity, material density, and elasticity of the body of the planet.

As the Earth spins about its axis, material near the planet's Equator must travel farther to make one rotation than material near to the spin axis must travel. All material has inertia; the tendency of a moving mass to continue moving in a straight line until stopped by an external force. This property makes the fast-moving material near the Equator to want to fly off from the planet in a straight line. The rest of the mass of the planet gravitationally attracts the material and keeps it glued to the planet, but the material's inertia makes the planet to bulge out at the Equator. Every geological layer is eventually and constantly bulged. The bulge that we observe on the surface is repeated at every geological layer and in various degrees of eccentricity depending on the layer mass and distance from the centre of the Earth. As the Mantle has double the density than the Crust and has bigger volume, we expect it to be highly bulged. This allows the Outer Core, which is a fivefold denser viscous liquid material than the Crust, to appear much oblate. In all situations the bulge major axis is aligned with the equatorial axis.

Geologic layers of the Earth

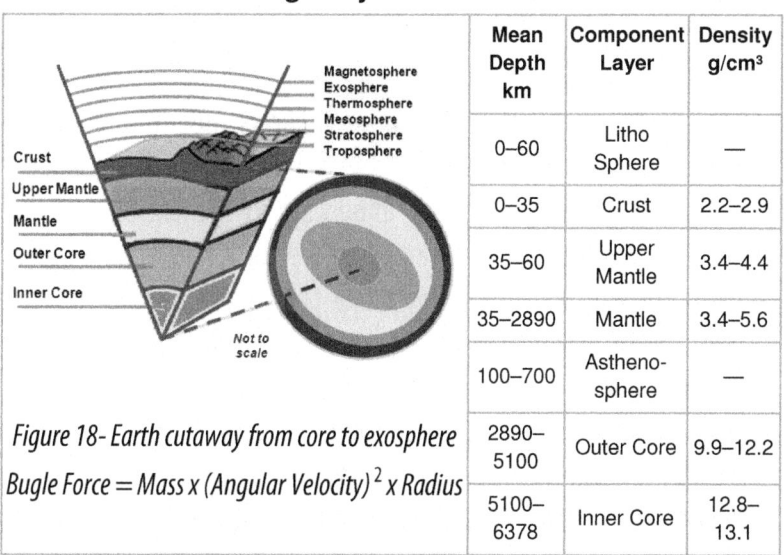

Mean Depth km	Component Layer	Density g/cm³
0–60	Litho Sphere	—
0–35	Crust	2.2–2.9
35–60	Upper Mantle	3.4–4.4
35–2890	Mantle	3.4–5.6
100–700	Asthenosphere	—
2890–5100	Outer Core	9.9–12.2
5100–6378	Inner Core	12.8–13.1

Figure 18- Earth cutaway from core to exosphere

Bugle Force = Mass x (Angular Velocity)² x Radius

The Crust is separated from the Mantle by the Mohorovičić discontinuity principle, and the thickness of the Crust varies; it averages 6 km (4 miles) under the oceans and 30-50 km (18.7-31.3 miles) on the continents. The internal thermal energy of the planet (Sanders, 2003) is probably produced by the radioactive decay of potassium, uranium and thorium isotopic Inner Core. The tilting of the solid Inner Core inside the viscous Outer Core, does not mean that the planet Crust, more than 5,100 km (3,187 miles) above, will also tilt. Earthquakes and volcanoes are expected as a result, but not a Crust matching tilt. However, Earth's geography will change if Earth slows down its spin as bulged oceans deflate between the topics and rise over planet's north and south.

The atmospheric pressure at sea level is defined as 1 bar, with a scale height of about 8.5 km (5.3 miles). The atmosphere is 78 percent nitrogen and 21 percent oxygen, with trace amounts of water vapour, carbon dioxide and other gaseous molecules. Three-quarters of the atmosphere's mass (Moran, 2005) is contained within the lowest layer that is called the Troposphere. The height of the Troposphere varies with latitude, ranging between 8 km (5 miles) at the poles and 17 km (10.6 miles) at the Equator due to the bulge of the atmosphere layer similar to the bulge of each geological layer as introduced earlier. Other atmospheric functions that are important to life on Earth include transporting water vapour, providing useful gases, causing small meteors to burn up before they strike the surface. Trace molecules within the atmosphere serve to capture thermal energy, thereby keeping an average temperature and causing the so called greenhouse effect (Pidwirny, 2006). Carbon dioxide, water vapour, Methane and Ozone are the primary greenhouse gases in the Earth's atmosphere. Without this thermal energy-retention effect, the average surface temperature would be −18° Celsius (0° Fahrenheit) and life would likely not exist the way we know it. The Earth's atmosphere has no definite boundary. Rather, it gradually becomes thinner and fades into outer space. Ocean currents are also important factors in determining climate, particularly the Thermohaline circulation that distributes thermal energy from the equatorial oceans to the Polar Regions.

Some believe that the Ozone layer stands firm against the charged particles reaching Earth from the Sun. As manifested earlier, the true shield against such charged particles is the magnetic field that engulfs the Earth. Where it reaches highest intensity at the magnetic poles, a less thermal energy is generated when the slowed down short spiral-radius travelling protons collide with the thermosphere molecules and vice versa where the magnetic field intensity is low at the Thermosphere layer above the magnetic equator. The hot lower-density air at the hot regions rises and is replaced by cooler, higher-density air. The result is atmospheric circulation that drives the wind and climate through redistribution of thermal energy. The primary atmospheric circulation bands consist of the trade winds in the equatorial region below 30° latitude and the westerlies, in the mid-latitudes, between 30° and 60° latitude. Wind can also be generated due the difference of atmospheric pressure. If a large meteoroid passes close to Earth at super speed, it may cause a drop in air pressure at the point of close proximity to the Earth's atmosphere. This causes a plume of air, that rushes to fill the pressure drop, be formed. Such a drastic scenario may cause a colossal wind of epic proportions that could impact the whole surface of the Earth.

Water vapour generated through surface evaporation is transported by circulatory patterns in the atmosphere. When atmospheric conditions permit, an uplift of warm, humid air takes place. This water vapour condenses and drops to the surface as precipitation. Most of the water is then transported back to lower elevations by river networks. Usually, it returns to the oceans or is deposited into lakes and aquifers. This water cycle is a vital mechanism for supporting life on land and is a primary factor in the erosion of surface features over geological periods. Precipitation patterns vary widely, ranging from several meters of water per year to less than 100 millimeters. The melting of the ice caps results in extra water directly discharged into the oceans. The vaporization has therefore increased and, accordingly, so has the rate of precipitation. Atmospheric circulation, topological features and temperature differences determine the average precipitation that falls in each region. The Earth has been subdivided into specific latitudinal belts of approximately homogeneous climates.

The commonly-used Köppen climate classification system has broad groups of temperature belts (tropical, dry climate, middle latitude, continental and cold polar), which are further divided into more specific subtypes. Following the Thermosphere temperature, upon the shifting of the magnetic poles, all temperature belts will tilt. The combination of increased precipitation rate and the tilting of temperature belts would tell which regions to experience drought and which to get flooded. As a result, a new precipitation map is created. Climate can also be classified based on the temperature and precipitation, with the climate regions characterized by fairly uniform air masses. The upper atmosphere above Troposphere is usually divided into the Stratosphere, Mesosphere, and Thermosphere. Beyond these, the Exosphere thins out into the Magnetosphere. The Kármán line, defined as 100 km (62.5 miles) above the Earth's surface, is a working definition for the boundary at which the low density atmosphere layer begins for 700 km (438 miles). Since the boundary is not a solid one, there is a slow but steady leakage of the atmosphere into space. Because unfixed hydrogen has a low molecular weight, it can achieve escape velocity more readily, and it leaks into outer space at a greater rate than other gases. For this reason, the Earth's current environment is oxidizing, rather than reducing. The loss of reducing agents such as hydrogen is believed to have been a necessary precondition for the widespread accumulation of oxygen in the atmosphere (David C.

Catling, 2001) The oxygen-rich atmosphere also preserves much of the surviving hydrogen by locking it up in water molecules.

The magnetic field results from magnetic monopoles. As its name implies, a magnetic monopole has only one magnetic pole (either a north pole or a south pole). In other words, it would possess a "magnetic charge" analogous to electric charge (Boothroyd, 2009). It is also found that a magnetic monopole is a stream of photons oscillating at a specific frequency (Ng, 2002). The magnetic monopoles of a bar magnet, for example, result from the extra spin of electrons, which point in one direction, than from the electrons that point in the opposite direction of electron pairs that lie in the optimal geometrical arrangement of the metal. Roger Penrose explains in his book, *The Emperor's New Mind* that the momentum wave function formula, that tells the time and position of a photon, is a corkscrew or helix (Penrose & Gardner, 2002). Thus one can visualize magnetic attraction and repulsion between two magnets as streams of photons with their corkscrew shaped wave functions screwing into 'attraction' or screwing out of 'repulsion' with each other. As forces do indeed arise from an exchange of particles, it has been suggested that such streams of photons travel in what we call the magnetic force lines. He also explains that all emitted photons must carry some mass because according to Einstein's formula $E=mc^2$ where 'E' is the energy, 'm' is the mass and 'c' is the speed of light. As demonstrated earlier in chapter one, Photons exhibit wave-particle duality; i.e. they enjoy and own both wave-like and particle-like properties. Such a wave-particle duality is a central concept of Quantum Physics; the branch of physics that explains the interactions of energy and matter. This duality addresses the inadequacy of classical concepts like "particle" and "wave" in fully describing the behavior of quantum-scale atoms and smaller elements such as photons. A Planck's formula applies at such minute scale. The formula states that the energy of a travelling photon is proportionate to its frequency E=hf where 'E' is the energy, 'h' is Planck's constant and 'f' is the photon frequency. Not only because they were colleagues and friends but by putting both Einstein and Planck's formulae together, we find that the mass of a photon is directly proportionate to its frequency or $E=mc^2=hf$.

This means that the higher the frequency of a photon gets, the higher its mass becomes. Since mass in motion creates force, magnetic force is hence produced following Newton second law of motion; F=ma, where 'F' is the force, 'm' is the mass and 'a' is the rate of change of particle velocity with time. Such a force that is established between any two magnets could, therefore, lead to a force that is strong enough to tilt one or the two magnets off their original positions. **So, what is the source of Earth magnetic field, one may wonder?** As explained earlier, the Inner Core is isolated from the rest of Earth by the low-viscosity fluid high density Outer Core. The Inner Core can spin, nod, wobble, oscillate, and even flip over, being loosely constrained by the surrounding Outer Core shell. The radioactive decay of the Inner Core isotopes creates thermal energy and swamps of free electrons that are transferred by convection through the Outer Core to the Mantle. Due to the bulge of the Outer Core, the emitted electrons find it easier to flow along the major axis of the oblate protruded Outer Core. The electronic current could therefore be named the "Equatorial Electric Current". A magnetic field is thus induced and is constantly maintained by the flow of the swirling electrons in the Outer Core. A known requirement for the induction of a magnetic field is a rotating fluid. Particles moving in a straight line inside a rotating fluid would seem to be moving in circles when observed from an external non-moving frame of reference. Newton's laws of motion govern the motion of an object in an inertial frame of reference, in which matter remains at rest or stay in motion unless acted upon by an outside force. However, when transforming Newton's laws to a rotating frame of reference, the Coriolis force appears, together with the centrifugal force.

The Coriolis force acts in a direction perpendicular to the rotation axis. The force is proportionate to the object's speed in the rotating frame. The Coriolis force tends to organize fluid motions. It organizes the electrons currents into circular motions around columns that are parallel to the rotation axis of the rotating frame. Electronic current in any circuit produces magnetic field that is perpendicular to the plane of the electric path. Consequently, magnetic force lines, in centres of circular-helical motions of the electronic currents should be parallel to the axial rotation of the Earth. As explained earlier, there is, currently, a 10° gap angle between the Earth's axis of rotation and the imaginary line connecting the magnetic poles. This gap angle is changing as manifested by the wandering poles. If the magnetic field that we observe on the surface of the Earth arouse from the induced magnetic field force of the free swirling electrons that flow through the Outer Core, as most Scientists claim, then we should expect to observe no gap between the axis connecting the two magnetic poles, and the axis of the rotation of the Earth! So, why then does such a 10° gap angle exist?

 Moving objects on the surface of the Earth experience a Coriolis force, and appear to veer to the right in the northern hemisphere, and to the left in the southern hemisphere. Exactly on the Equator, motion east or west, remains precariously along the line of the Equator. Movements of air in the atmosphere and whirlpools in the ocean are notable examples of such a behavior. Rather than flowing directly into the sink, water flows in a counterclockwise direction in the northern hemisphere and in a clockwise direction in the southern hemisphere.

The Coriolis force is responsible for the direction of rotation of large cyclones. In electrical terms, an electric current always flows in an opposite direction to the flow direction of its electrons. Upon applying Coriolis force onto the electric currents that move counterclockwise in the lower half of the Outer Core, a magnetic field is induced in the opposite polarity to the magnetic field that is induced by the clockwise movement of electric current in the upper half of the Outer Core; as shown in Figure 19.

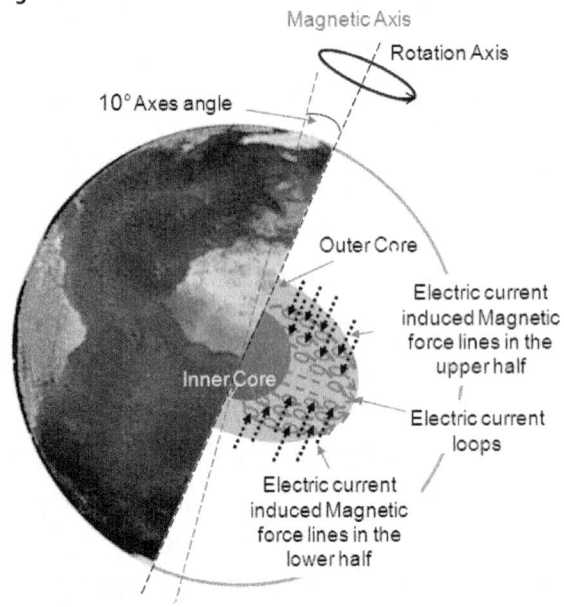

Figure 19- Inner Core and Outer Core magnetic fields are not aligned

As the speed of the electric current varies along its 2,200 km (1,367 miles) path from the Inner Core to the Inner Mantle, the strength of the induced magnetic force lines will also vary and will mostly be observed on the Earth's surface around the geographical poles at the surface projection area of the Outer Core beneath. The contours shown in Figure 20 are the time rate of change of the magnetic strength in the North and South Poles (GIRIJA RAJARAM, 2002). Full curves show contours where the rate of change of magnetic flux is positive while dotted contours refer to regions where the rate of change is negative. These latter contours are found to partly overlap with regions of Reverse Magnetic Flux where the radial field points in a direction opposite to that expected for a dipole magnetic field. The contours of negative rate of change are seen to straddle almost the whole of the Antarctic continental and large expanses of the adjoining ocean areas below South America and South Africa. There remain strong flux points that define the normal north and south magnetic poles as we observe on a typical compass. The research by the scholars of the Indian Institute of Geomagnetism, Colaba, Mumbai, India goes on to describe four static flux bundles of intense magnetic flux, the two northern ones being located below Arctic Canada and Siberia, and the two southern ones located below Antarctica. This feature makes it impossible for the Earth to have one dipole magnet source no matter what origin it has. These

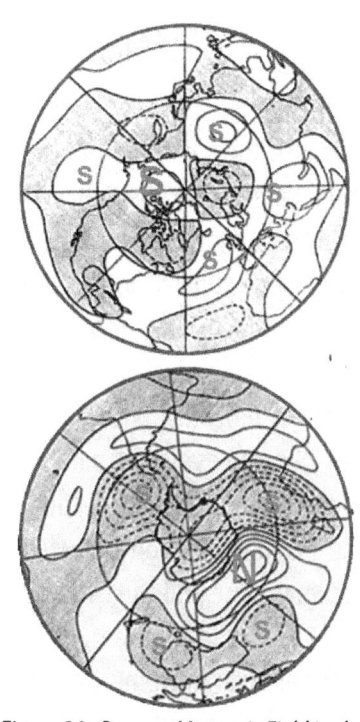

Figure 20- Reverse Magnetic Field in the Antarctic

four flux bundles have remained almost stationary for over 250 years. It also describes that over the period 1842 AD to 1980 AD when the magnetic dipole field decayed by 7%, regions of Reverse Magnetic Flux created patches that have increased in size and in intensity. The observation suggests the existence of more than one magnet and that some are riding over in the weakness of others. There remains one possibility for the source of the single strong magnetic field, that distinguish itself on the surface and around the Earth; and that is the Inner Core. Figure 21 illustrates a three magnet configuration that can well explain the mystery of the source of Earth's magnetism and the source of the spinning of the planet. The two cylindrically shaped mono pole magnets are induced by the swirling free electrons flowing in the Outer Core and the third is a strong dipole magnet that originates from the Inner Core. While the upper semi-sphere Outer Core magnetic force lines add to the Inner Core dipole magnet and create a stronger *pull* force, the lower semi-sphere Outer Core magnetic force lines repel the Inner Core dipole magnet and create a repelling *push* force. The two forces act in parallel and are in the same direction. The resultant friction between the Inner Core and the rotating Outer Core leads to the transfer of the angular movement from

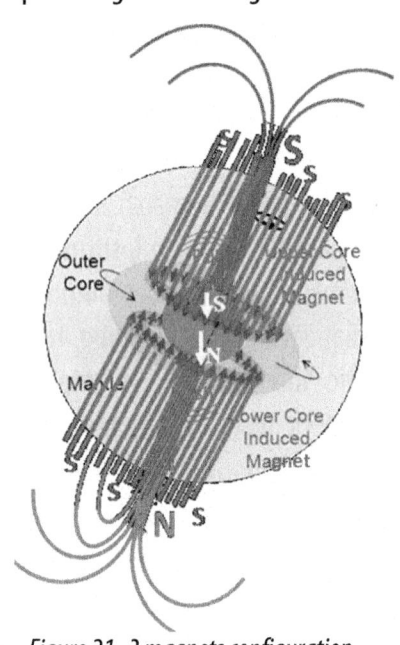

Figure 21- 3 magnets configuration

the Outer Core to the Inner Core. This keeps the Inner Core spinning at the same angular velocity as the rest of the Earth. **However, a question jumps to the front; what brings the planet to spin in the first place?**

What makes Earth spin? We could remember, while studying at high school, that a charged particle that moves into a magnetic field experiences a sideway force. Such a force is proportionate to the strength of the charge, the speed of the particle, and the strength of the component of the magnetic field that is perpendicular to the path of the charged particle. Such a rule is behind the basic theory of the Motor engine and is called the Left- Hand rule. The rule states that the direction of the force, Lorentz force, is in line with the direction of the left hand thumb so long as the direction of the magnetic field is aligned with the direction of the index finger and that the direction of the electric current is aligned with the rest of the left hand fingers as shown in Figure 22, respectively perpendicular to one another.

The magnetic field fluxing out of the N pole of the Inner Core towards its S pole exerts a magnetic force pointing almost toward geographic North. The negatively charged electrons or as we called them earlier "Equatorial Electric Current"

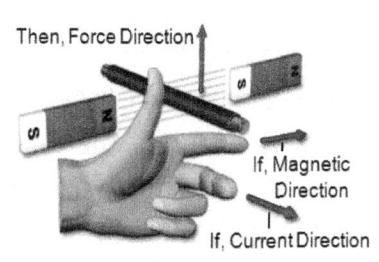

Figure 22- Left Hand Rule

flowing, in direction terms, onto the major plane of the bulged Outer Core towards the Inner Mantle means, in electric notation terms, positive electric currents flowing in the opposite direction i.e. towards the centre of the Earth. Applying the Left-Hand rule on two opposite sides of the equatorial perimeter of the Outer Core as shown on Figure 24, demonstrates that Lorentz forces are created of pairs of almost equal magnitude and opposite directions along the perimeter of the bulged Outer Core. The perimeter torque of such pairs of forces keeps the spinning of the planet going counterclockwise, for the moment.

The Earth achieves one complete spin around its axis every 86,400 seconds. **Hence, on may wonder, what controls the speed of the spinning?** And what happens if the lines of the Inner Core magnetic field were not at a right angle with the electric current? Or, that the angle between the Inner Core magnetic field and the rotational axis is changed? The science of trigonometry can help answer these questions. When a force is acting by an angle to the horizontal direction, it could be segmented into horizontal and vertical components. The vertical component is a fraction of the main force and is calculated by multiplying the force by the "cosine" of the angle that falls between the direction of the force and the vertical direction. You could tell from Figure 23 that when the angle is small the vertical component of the force is bigger than when the angle is large. The rotation axis of the Earth is perpendicular to the direction of "Equatorial Electric Current". Since the current gap angle between the magnetic force and the rotation axis is 10°, the vertical component of the magnetic force is therefore calculated by multiplying the magnetic force by cosine 10°. As the Inner Core is free to tilt, swivel and rotate, its alignment with respect to the vertical direction or rotation axis may vary. The vertical component of the magnetic force will increase and decrease depending on such variations. The Lorentz force which is proportionate to the strengths of the electric current and the vertical component of the magnetic force will vary accordingly. The lesser the force, the lesser the torques that rotate the planet, and accordingly the slower the planet will spin. The rate of Earth's spinning increases and decreases across rather lengthy stretches of time. Spinning observations are tabled for some 374 years (from the year 1623 to 1997) (IERS, 2000).

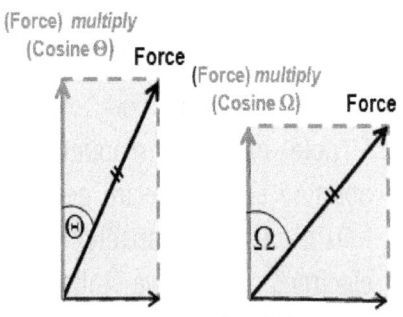

Figure 23- Same Force but different gap angle results in different force components on the vertical axis direction=(Force) multiply (Cosine of the gap angle)

Throughout the many years, it is apparent that the length of the solar-day was a fraction of a second faster than 86,400 seconds about 41 percent of the time. It is equally apparent that the length of the solar-day was a fraction of a second slower than 86,400 seconds about 59 percent of the time. It merits further research of this matter to inspect the association of time records of variation of gap angle of the wandering of the magnetic poles versus the spinning speed of the planet.

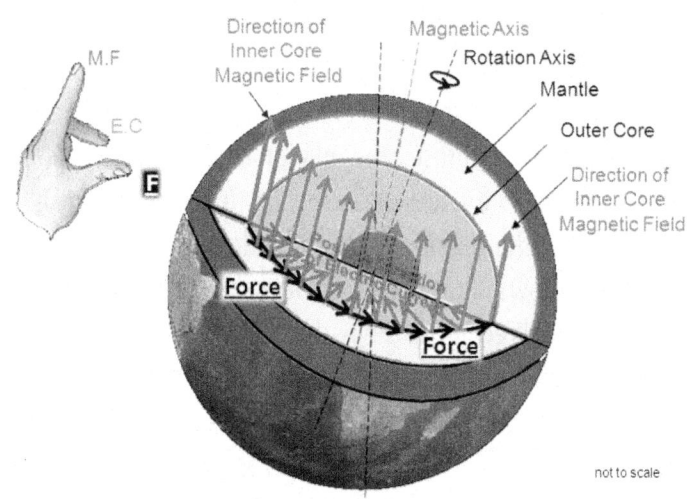

Figure 24- The torque generated by the Left-Hand rule forces makes Earth spin

Unlike what is currently documented in most text books and based on the above information, it is deduced that the magnetic field that we observe and measure on the surface of the planet is emanating from the Inner and not the Outer Core. It is also deuced that the electrons emanating from the Inner Core, through the bulged Outer Core, create electric currents or "Equatorial Electric Current" onto the equatorial plane. Such electric currents act together with the magnetic field components that are perpendicular to its directions to generate driving forces that form pairs of torque, along the perimeter of the Inner

Mantle, that keep the planet spinning about its axis as shown in Figure 24. But, what if another celestial body that possesses a strong magnetism comes in close vicinity to the magnetically balanced solar system? The dipole magnet of the Inner Core of the Earth is thus subjected to a new and additional magnetic force emerging from the intruding planet. The Inner Core dipole magnet will be influenced by the resultant magnetic field configuration and the Inner Core may tilt. The spinning of the Earth will be impacted when the Inner Core starts to tilt and the gap angle changes. If the gap angle gets larger, the Earth rotation will slow down. The Lorentz force would die out when the gap angle reaches a 90º and the Earth will cease to spin about its axis. If the gap angle becomes obtuse, the planet could start to spin in the opposite direction, i.e. clockwise. The Earth's axis of spin is always perpendicular to the major axis of the bulged Outer Core where the electric current flows. Most planets in our solar system, including Earth, spin in the same direction as they orbit the Sun. The spinning axes are nearly vertical to the solar system plateau. The exceptions are Uranus and Venus where the former spins nearly horizontal relative to its orbit. The bulge around the Equator of Uranus is about 2 percent of the radius, or about 500 km (300 miles). As we explore historical and holy citations in the third chapter of this book, we shall be able to establish the time at which incidents of core tilt took place and the time cycle of repetition.

The magnetic field forms the Magnetosphere, which traps the charged particles of the solar mass energy. The sunward edge of the bow shock is located at about thirteen times the radius of the Earth. The Sun dissipates two types of energies: the feeble radiant energy as in light i.e. photons and the mass energy in particles such as protons, neutrons and electrons. The interaction between the magnetic field and the charged particles forms the Van Allen radiation belts; a pair of concentric, horn-shaped zones of radiation containing electrons and ions and extends thousands of kilometers into space; the belts were named after their discoverer, James Van Allen (1914–2006). The electrons in the Van Allen belts are normally not dangerous to satellites and spacecraft, but every month or so radiation levels spike to as much as a thousand times their usual intensity. These surges, called Geomagnetic Storms, are related to increased intensity of the solar wind. Recently, leaks have been detected in the magnetic field which interacts with the solar wind in a manner opposite to that of the original hypothesis (Thompson, 2008).

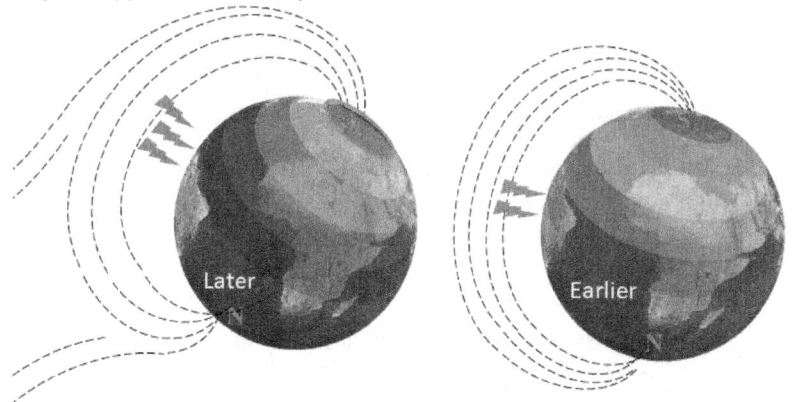

Figure 25- Ice caps & temperature belts will follow the tilted magnetic poles

During solar storms, this could result in large-scale blackouts and disruptions in artificial satellites. Van Allen internal belts lie between 100 and 700 km (60 and 420 miles) from the surface of the Earth and are composed of the protons of the mass energy arriving from the Sun and other cosmic sources. Such protons can not escape the magnetic field of the Earth and as explained earlier they contribute, due to collision with air molecules, to a variable thermal energy so that the "magnetic equator" will always maintain the highest temperature on the surface of the planet. This also could explain why we find ice caps at the two highly intense magnetic spots on the planet. When the magnetic poles move away from their current positions at the geographic poles, the incumbent ice caps become unshielded and start to melt. The new ice caps will form wherever the magnetic pole will finally settle, causing an entirely new alignment of the temperature belts and new precipitation map as shown in Figure 25 and resulting in **Climate Exchange** where some regions will become warmer and other regions become colder.

Position and angle variations of the Earth and the Moon make a great difference on life on Earth. The Moon revolves around the Earth every 27.32 days relative to the stars. When combined with the Earth revolution about itself and around the Sun, the period of the synodic month, from new Moon to new Moon, is 29.53 days (Williams, 2004). Viewed from the celestial North Pole, the motion of Earth, the Moon and their axial rotations are all counterclockwise. Viewed from a vantage point above the north poles of both the Sun and the Earth, the Earth appears to revolve in a counterclockwise direction around the Sun. The orbital and axial planes are not precisely aligned: Earth's axis is tilted some 23.44° from the perpendicular to the Earth–Sun plane and the Earth–Moon plane is tilted about 5° against the Earth-Sun plane. Without this tilt, there would be an eclipse every two weeks, alternating between lunar eclipses and solar eclipses. Because of the axial tilt of the Earth, the position of the Sun in the sky as seen by an observer on the surface varies over the course of the year. For an observer at northern latitude, when the northern pole is tilted toward the Sun the day lasts longer and the Sun appears to climb higher in the sky. This results in warmer average temperatures from longer exposure to the collision of charged particles at the Thermosphere layer and consequent thermal energy reaching the Earth's surface. When the northern pole is tilted away from the Sun, the reverse is true and the climate is generally colder. Above the Arctic Circle, an extreme case is reached where there is no daylight at all for part of the year. This is called a polar night. This variation in the climate because of the direction of the Earth's axial tilt, results in the seasons. By astronomical convention, the four seasons are determined by the solstices (the points in the orbit of maximum axial tilt toward or away from the Sun) and the equinoxes (the points at which the direction of the tilt and the direction to the Sun are perpendicular). Winter Solstice occurs about December 21, Summer Solstice is near June 21, Spring Equinox is around March 20 and Autumnal Equinox is on September 23. The angle of the Earth's tilt is relatively stable over long periods of time. The Orientation, rather than the angle, of the Earth's axis is believed to change over time, processing around in a complete circle over each 25,800-year cycle, which is the same period it takes the Sun to complete one full cycle about its axis. This means that 12,900 years ago and in opposite connotation to current status; the winter season started on June 21st and not December 21st in the Northern Hemisphere and vice versa in the Southern Hemisphere. A spinning top toy could bring us closer to understand the driver of Earth experiencing such a Precession Cycle. Figure 27 shows two forces that work

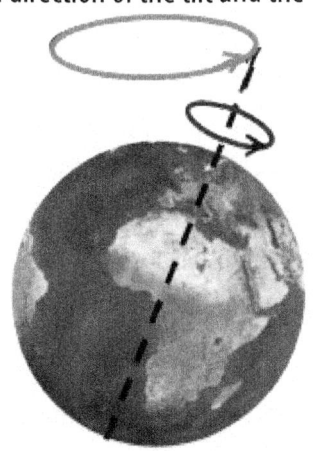

Figure 26- A full circle of ecliptic precession takes 25,800 years.

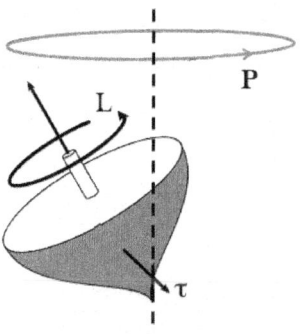

Figure 27-Forces applied on a spinning top toy

together to cause a Precession Cycle 'P'. They are the Angular Velocity 'L' and the Angular Torque 'τ'. It is remarkable to notice that the spinning top toy will never flip so long as the torque 'τ' is at right angle from the axis of its rotation about itself. By analogy we have similar kind of forces acting on Earth; where the angular velocity is a product of pairs of similar but opposite forces applied along the equatorial plane as explained earlier with the help of the 'Left Hand Rule', and the orbital force that keeps Earth onto its path orbiting the Sun. Earth rotates around the Sun at a determined distance thanks to the equilibrium between the Sun gravitational force that pulls Earth towards it and the Earth centrifugal force that pulls the Earth away from the centre of its orbit. In modern times the Earth's Perihelion—the point of closet approach to the Sun—occurs around January 3, and the Aphelion, the point when it is farthest from the Sun, around July 4.

However, these dates change over time due to Precession[13] or change in the direction of the axis of the Earth. The changing Earth-Sun distance results in an increase of about 6.9 percent in mass energy reaching the Earth at Perihelion relative to Aphelion. Since the Southern Hemisphere is tilted toward the Sun at about the same time that the Earth reaches the closest approach to the Sun, the Southern Hemisphere receives slightly more radiant energy from the Sun and the Thermosphere than does the Northern Hemisphere over the course of a year. This effect is much less significant than the total thermal energy arriving from the Thermosphere layer which is much dependent of the magnetic field map and not the axial tilt of the planet.

The tilting pole is not a new phenomenon. A hippo-like mammal known as Coryphodon was one of several ancient mammal groups that

---

[13] http://en.wikipedia.org/wiki/Precession_(astronomy)#Astronomy

lived in the high Arctic fifty three million years ago, according to a new study led by the University of Colorado at Boulder. A tropical belt that passed the celestial or geographic poles must have otherwise been in place. Ancestors of tapirs and ancient cousins of rhinos living above the Arctic Circle endured six months of darkness each year in a far milder climate than today's. It featured lush, swampy forests, according to Jaelyn Eberle, assistant Professor, department of Geological Sciences and Curator of Vertebrate Paleontology, University of Colorado Museum and chief study author.

The study [14] suggests several varieties of prehistoric mammals as heavy as 450 kg (1,000 pounds) each lived on what is today Ellesmere Island, near Greenland, on a summer diet of flowering plants, deciduous

Figure 28-Coryphodon

leaves and aquatic vegetation. But in winter, they apparently switched over to foods like twigs, leaf litter, evergreen needles, and fungi, according to Eberle. The study shows the implications for the dispersal of early mammals across polar land bridges into North America and for modern mammals that likely will begin moving north if Earth's climate continues to change. Henry Fricke of Colorado College in Colorado Springs and John Humphrey of The Colorado School of Mines in Golden co- authored a paper on this subject. The team used an analysis of carbon and oxygen isotopes extracted from the fossil teeth of three varieties of mammals from Ellesmere Island: a Coryphodon, a second, smaller ancestor of today's tapirs; and a third, rhino- like mammal known as Brontothere. "Animal teeth are among the most valuable

---

[14] http://www.physorg.com/news163081573.html; 53 million-year-old high Arctic mammals wintered in darkness

fossils in the high Arctic because they are extremely hard and better able to survive the harsh freeze-thaw cycle that occurs each year", according to Eberle. Telltale isotopic signatures of carbon from enamel layers that form sequentially during tooth eruption allowed the team to pinpoint the types of plant materials consumed by the mammals as they ate their way across the landscape through the seasons.

Key among the most recent findings is the discovery that the current structure of the ecological community is based on climate changes that took place thousands of years ago. Ice shelves are permanent floating ice sheets that are attached to land and are constantly fed by glaciers. Glaciers are enormous masses of terrestrial ice that form on land regions through the compaction and re-crystallization of snow into ice. The largest ice shelf is the Ross Ice Shelf in Antarctica, which ranges from 180 to 900 m (600 to 3000 ft) in thickness and is about 960 km (600 miles) long, the size of France or Sudan's Darfur. The cliffs at the water edge are about 60 m (200 ft) high. "The current ecosystem depends, in a very direct way, on the legacy of past climatic effects", said Andrew G. Fountain Professor of Geography and Geology at Oregon's Portland State University. For example, researchers have determined that about 40,000 years ago the Ross Ice Shelf had advanced and blocked the Taylor Valley's opening to the ocean. A huge lake subsequently filled the valley, and algal mats formed at the lake bottom. About 8,000 years ago, the Ross Ice Shelf retracted and the lake drained out, leaving the algal mats behind. "Today, that relic algal mat is thought to be the major carbon source for the soil ecosystems", according to Fountain."The spatial dimensions of the abundance and diversity of the soil invertebrates is partly determined on this relic carbon source". John Priscu, Professor of Ecology at Montana State University in Bozeman who has conducted research in Antarctica since the early 1980s, said that the McMurdo Dry Valleys, a cold, barren desert comprising 4,800 square km (1,853 square miles) in southeastern Antarctica are "one of the oldest landscapes on the planet". It is the frozen continent's largest ice-free area and has become a focal point of scientific research in the 100 years since its discovery[15]. Priscu found out the McMurdo Dry Valleys have actually cooled off slightly in the past thirty-five years.

As discussed earlier, the magnetic poles shift would weaken the current ice caps of their strong magnetic shield thus causing the ice to melt and the sea level to rise. We are currently experiencing the first stage of such an ice meltdown; it is gradual. The rising sea level is offset by an increase in the water evaporation rate. Nowadays the airborne water has contributed to more powerful hurricanes and gale winds especially in south east Asia where the lands receive the run up rain clouds that emanated in abundance as a result of the increased evaporation at the Pacific Ocean, the largest on the planet. We also observe increased volume of precipitation along the Equator line where the vaporization is at its peak due to the weakness of the magnetic shield that results as the magnetic force lines are deflated towards the intruding planet. The weakness of the magnetic field leads to the Van Allen inner belt protons gaining higher speed and longer spiral-radius motion around the magnetic force lines. The collisions between such spiraling protons and Thermosphere molecules become fiercer, leading to increased thermal energy reaching the Troposphere and causing global warming. While the first stage of ice meltdown is gradual, the second stage, as we shall discuss on the following chapter, could be abrupt. The sudden meltdown of the rest of the ice caps, both north and south, will create a sudden increase in sea level with more vaporization and more of

---

[15] Antarctic Desert Rich With Insights Into Life on the Edge, National Geographic

continuous rain until the sea has returned to its previous level. As the core tilt settles down, the magnetic shields at the new north and south magnetic poles would help in the formation of two new ice caps. No one could tell for sure the new locations of the magnetic poles and hence the new ice caps.

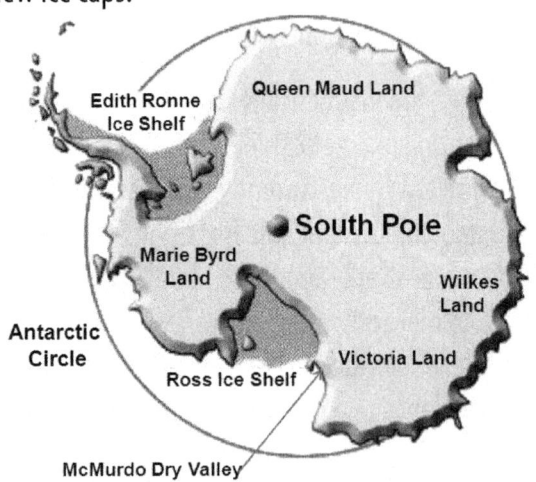

Figure 29- Ross Ice Shelf retracting

It could be possible that they follow the same path that resulted in a tropical belt passing through Elsmere Island as discussed earlier! In this case the ice caps new locations could exist in India and Brazil. Consequently, the temperature belts are tilted almost 70° to 80° from current locations.

People normally do not carry oracles' prophecies seriously. But, could Nostradamus, through meditation and remote viewing as we explained in the first chapter of this book, be correct in the prediction that he made almost 500 years ago? In fear of prosecution at the time, Nostradamus has arranged his predictions in concealed quartets of poetry. Figure 30 brings his quartet which, upon reading, one cannot stop wondering. Will famine emerge as a result of heat waves that overwhelm the whole planet as in global warming's? Does it tell that the ice caps will totally melt and that the molten ice would result in excess water of the molten ice volume?

> Si grand famine par onde pestifère.
> Par pluie longue le long du pole arctique,
> Samarobrin cent lieux de l'hémisphère,
> Vivront sans loi exempt de politique.
>
> Very great famine through pestiferous wave,
> Through long rain the length of the arctic pole:
> Samarobrin one hundred leagues from the hemisphere,
> They will live without law exempt from politics.

Figure 30- Prophecies, Century VI, Quatrain 5

Does it tell us that vaporization will take over and that rain will continue until the sea loses such a surplus of water? Is that "Samarobrin" a name of a man, a machine, an organization or some other thing? One cannot ignore the fact that the word includes the same letters as Obama, the president of the United States. Could it be that **Obama** will sound the **siren** (sirène) and alarm at the countries to pull their act together against the changes ahead?

A new route is breaking the ice, as a result of the gradual disappearance of the current ice cap in the north. A whole-year-round ship route between Asia and Europe is being naturally created. No longer will ships, going from Denmark to Japan, for instance, have to pass through the Panama Canal or Suez Canal. It would be much more direct to cut through the waters of the north. A time-saving new shipping route connecting Europe to the Pacific may be navigable thanks to the melting of the ice cap in the north, scientists say. The Northern Sea Route, which cuts through the Arctic waters along Canada, was until 2007 blocked by ice. However, the steady rise in temperatures over the last twenty years has caused the ice to thin retreat substantially. It has shrunk by 50% over 27 years and by 30% during eight years only from the year 2000 to 2007. This means that the passage may soon become navigable for most of the year. Peter Wadhams Professor of Ocean Physics, and Head of the Polar Ocean Physics Group in the Department of Applied Mathematics and Theoretical Physics, University of Cambridge, UK said that the long-sought dream of a Northeast shipping route "could become reality[16] in as little as ten years' time". I believe that, given the aforementioned cause-and-effect chain, the process is accelerating even faster. Commercial companies trading by sea, between Europe and the Far East or northwestern America would see travel times slashed. At the moment, a voyage from London to Japan via the Suez Canal covering 20,300 km (11,000 nautical miles) takes thirty-five days. The distance via the Northern Sea Route is only 13,000 km (7,000 nautical miles) and would take around twenty-two days, a substantially shorter journey. Ships can currently reach the East through the Panama Canal or the Suez Canal, or by going around Africa or South America. It may not be a bad idea to re-assess the expansion projects at both canals, as traffic volume may not be as growing across these passages as it was contemplated earlier. As a result of tilting ice cap, the Northern Sea route on the Canadian side will be opened much faster though.

Mars temperature. Mars is the fourth planet from the Sun in the Solar System. Like Earth, Mars has a dense, metallic core region overlaid by less dense materials. Current models of the planet's interior imply a core region about 1794 ± 65 km (1121 ± 41 miles) in radius, consisting primarily of iron, nickel and sulfur. This iron sulfide core is partially fluid, and has twice the concentration of the lighter elements

---

[16] http://news.bbc.co.uk/2/hi/science/nature/705320.stm

that exist at Earth's core allowing induced magnets to emerge due to planet spin speed of 868 km/h (542 miles/h) at the equatorial level. The planet is tilted by 25.19° about the vertical axis to the solar plateau. As with the case of Earth, the tilting produces seasons cycles much like that on Earth. The lengths of the Martian seasons are about twice those of Earth's, as Mars's greater distance from the Sun leads to the Martian year being about two Earth years long. The Martian surface temperatures vary from lows of about −143° Celsius (−225° F) (at the winter polar caps) to highs of up to 35° Celsius (95° F) (in equatorial summer).

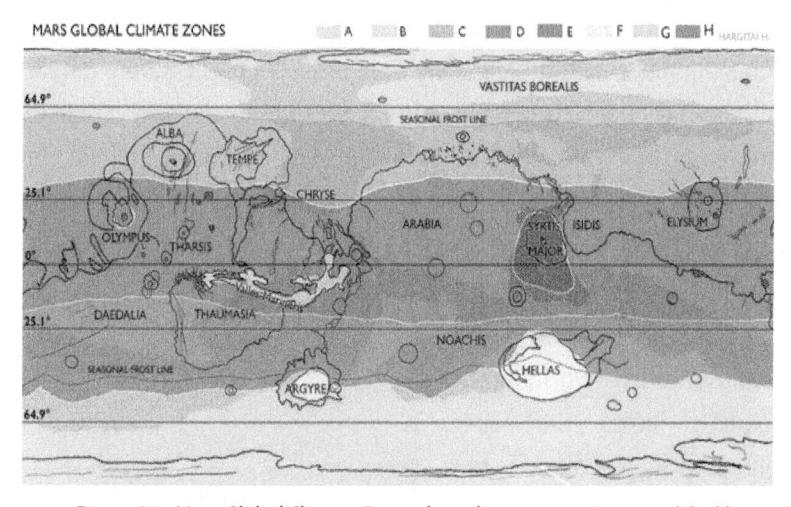

Figure 31- Mars Global Climate Zones, based on temperature, modified by topography, albedo, actual radiation.

- A=Glacial (permanent ice cap);
- B=Polar (covered by frost during the winter which sublimates during the summer);
- C=North (mild) Transitional (Ca) and
- C=South (extreme) Transitional (Cb);
- D= Tropical;

- E= Low albedo tropical;
- F= Sub polar Lowland (Basins);
- G=Tropical Lowland (Chasmata);
- H=Subtropical Highland (Mountain)

If the radiation energy of the Sun were directly responsible for the temperature variation on the surface of the planet, we should have expected the region of close proximity and at right angle projection to the radiation energy path to be the hottest! Yet the Martian temperature belts are tilted by 25.19° approximately than the region of close proximity to the Sun. They are much centered on the axis of the spinning of the planet. Scientists believe that Mars does not have the global magnetic field that, on Earth, is the source of the aurora borealis and the antipodal aurora australis. According to the physicists, the auroras on Mars are not due to a planet-wide magnetic field, but instead are associated with patches of strong magnetic field, primarily in the southern hemisphere which I tend to disagree and I will explain why. Earth has a magnetic configuration which is composed of A) two induced magnetic fields that are generated in the outer core. They are weak and appear in patches of south magnets in the southern and northern polar circles and B) one permanent magnet; the inner core, that is strongly fluxing out of Antarctica, circling around the planet and is observed as the Earth-wide magnetic field as explained earlier.

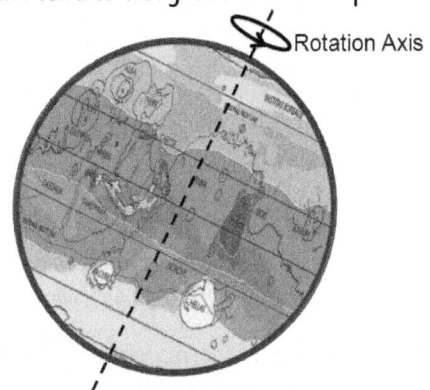

Figure 32- Mars Temperature Belts

Mars magnets setup. The caption released by NASA of the magnetic anomalies on its North Pole shows that the orientation and magnitude of the magnetic field measured by the Mars Global Surveyor's magnetometer as it sped over the surface of Mars during an early aero braking pass[17]. At each point along the spacecraft trajectory there drawn vectors in the direction of the magnetic field measured at that instant; the length of the line is scaled to show the relative magnitude of the field. As the magnetometer passed over the surface the needle swung rapidly, first pointing towards the planet (meaning a south magnet) and then rotating quickly towards up (meaning a north magnet) and back down again. All in a relatively short span of time, during which time the spacecraft has traveled a couple of hundred miles. Such changes of magnetic field can be explained if we consider that, similar to Earth, the Martian core has a 3-magnet configuration: A) two induced

Figure 33- Mars magnetic measurement

magnetic fields that are generated in the liquid part of the core. They are weak and appear in rings of patches of south magnets (Plasmoids) in the southern polar circle and northern polar circle and B) one permanent magnet; the solid inner part of the core, that is aligned with the axis of rotation of Mars and that has turned weaker and its flux out of the South Pole is consumed in three parts, as shown in Figure 34:

1. A relatively strong magnetic field force lines (A) that peek, along the rotation axis, through the planet geo-layers to reach its southern polar circle and loop back into the planet geo-layers, in parallel and

---

[17] http://photojournal.jpl.nasa.gov/catalog/pia00946

in close proximity to the rotation axis, to flux out at the northern polar circle. Such magnetic field force lines peek again through the planet geo-layer along the rotation axis to close the dipole north-south magnetic loop.

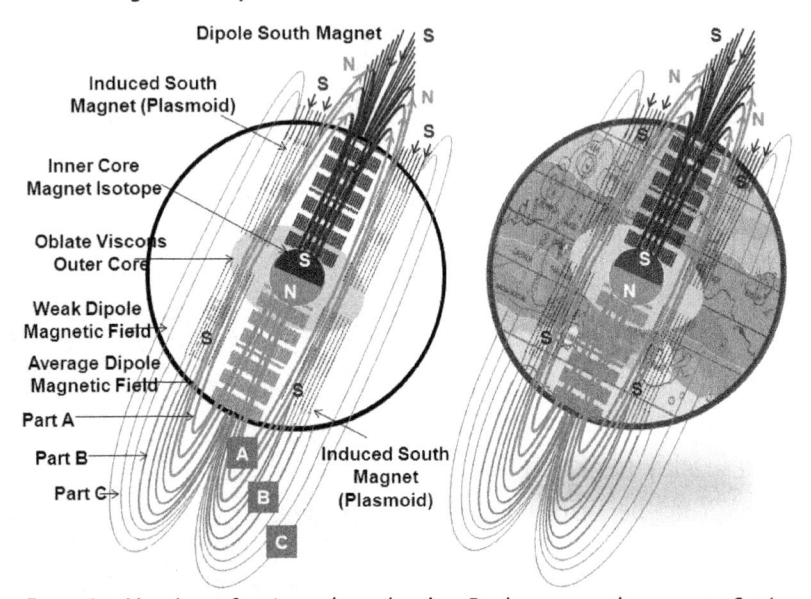

Figure 34- Mars Inner Core is much weaker than Earth; yet enough to support Sun' protons oscillating between its poles

2. A mild magnetic field force lines (B) that peek, along the rotation axis, through the planet geo-layers to reach its southern polar circle and loop back to the surface of the planet to connect with the ring of patches of induced south magnets (Plasmoids) in the southern hemisphere of Mars.
3. A weak magnetic field force lines (C) that peek, along the rotation axis, through the planet geo-layers to reach its southern polar circle and then loop back into Mars atmosphere to barely circumvent the planet surface from its South Pole to its North Pole.
4. This explains the magnetometer readings of south, followed by north and then followed by south magnetic field orientation while

travelling a couple of hundred miles over Mars' North Pole as shown in Figure 33.

Mars global warming. Such a configuration of magnetic field force lines stands against the charged particles arriving from the Sun at high kinetic energy levels and, similar to Earth thermosphere, causes the protons to spiral along the magnetic field force lines between the Martian magnetic poles. The weaker the magnetic field, the faster and longer spiral-radii motion will the protons pick up, leading to higher thermal energy upon collision between such protons and air molecules of the Martian atmosphere. This explains that at the magnetic equatorial level, where the magnetic field is weakest, the collision's thermal energy is at a maximum producing the highest temperature belt on the surface of the planet. Where the magnetic field is at its maximum strength, the protons will move at slower speed and shorter spiral-radii, producing aurora borealis and generating the minimum thermal energy upon mild collision with the Martian air molecules.

The approach of Tyche that is currently being investigated by NASA[18], to the inner Solar System, causes some of the magnetic field force lines emerging from the south pole of Mars to route into space instead of following one of the three paths explained earlier. This causes the magnetic intensity to drop allowing the protons to speed up and to spiral at longer radii around the magnetic field force lines; gaining more opportunity to collide with atmospheric molecules and to generate greater thermal energy. Polar ice starts to melt[19] and global warming is observed on Mars; much like what is happening on Earth.

---

[18] http://www.nasa.gov/mission_pages/WISE/news/wise20110218.html
[19] http://science.nasa.gov/science-news/science-at-nasa/2003/07aug_southpole/

Mars spinning speed. Needless to add, that Mars has similar spinning configuration as with Earth. Refer to: *What makes Earth spin* section. If the orientation of its permanent magnetic core is changed, the planet spin speed will change and may reverse depending on the resting angle between the magnetic field and the equatorial plane of the planet. While Mars has not shown yet any signs of slowing down, similar to Earth, other planets such as Venus and Saturn have started to show signs of slowing down. Cassini space craft took readings of the day-length indicator from Saturn. The results give 10 hours, 45 minutes, 45 seconds (plus or minus 36 seconds) as the length of time it takes Saturn to complete each rotation. That is about 6 minutes, or one percent, longer than the rotational period measured by the Voyager 1 and Voyager 2 spacecraft, which flew by Saturn in 1980 and 1981[20]. NASA scientists are looking for an explanation based on some variability in how the rotation deep inside Saturn drives the radio pulse.

I believe Saturn structure is similar to Earth and Mars and that the tilt of its magnetic core results in lesser Lorentz force to that drives the planet to spin so it slows down.

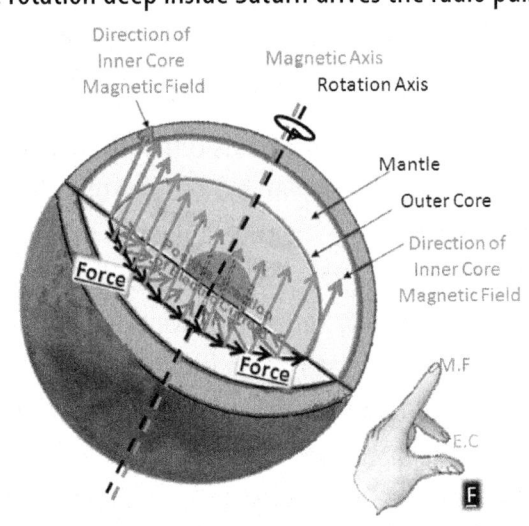

Figure 35- The torque generated by the Left-Hand rule forces makes Mars spin

---

[20] http://www.nasa.gov/mission_pages/cassini/media/cassini-062804.html

The Cycle

If climate change is attributed to the tilting of the core rather than anything else, could this be an isolated event? Or is it a more frequent one? And if it is more frequent, will the tilt be exactly the same in every occurrence? Could there be incidents of stronger swings that result in complete magnetic poles reversal? How did humans and other life forms on Earth respond to the consequential changes? And what do we learn that can help us to anticipate a new meteorological map? A decade ago, Djibouti and the Arabian Peninsula had less than 100 mm of rain per year. What if the rainfall rate in these areas suddenly jumps ten- or twenty fold? What will happen to the great lakes of Canada, Central Europe, Asia or Africa? How long will this new climate paradigm last? Looking back in history, it seems that every 3,550 years there is a pattern of Earth disorder: Earthquakes, volcanoes, climate change and more. The holy books have considerable information to share in this regard, but so do geological, historical and scientific knowledge.

The Exodus mystery was addressed (Wilson, 1985). The book drew attention to the astonishing similarities between the biblical calamity that took Egypt by surprise and the likely effects on Egypt of the Thera (Santorini) eruption. The Mediterranean Island blew up. A tsunami did virtually wipe out the Minoan civilization in Crete and south of Greece. Eight years ago, I visited outer rim which stands out as the final remains of the erupted island. Plato called it the site where Atlantis disappeared. The eruption, which is estimated to have taken place between 1500 and 1600 BC and is only 800 km (500 miles) away from the Egyptian coast, would have rattled the windows of modern Egypt.

The fallout cloud would have drifted high over Egypt and would have darkened the sky (Philips, 1998). According to Exodus 10:21-23

> "**21** And the LORD said unto Moses, Stretch out thine hand toward heaven, that there may be darkness over the land of Egypt, even darkness which may be felt. **22** And Moses stretched forth his hand toward heaven; and there was a thick darkness in all the land of Egypt three days: **23** They saw not one another, neither rose any from his place for three days: but all the children of Israel had light in their dwellings."

The event is attributed to a divine intervention. Graham Philips tried to establish the Thera eruption as the natural tool applied to fulfill God's plan. From a theological perspective, God created the forces of nature and they are HIS to use as HE pleases, so there is nothing unscientific or irreligious about attributing the plagues of Egypt to the eruption of Thera or to any other phenomenon. In Exodus 9:23-26 we are told that Egypt is afflicted by a terrible fiery hailstorm:

> "**23** And Moses stretched forth his rod toward heaven: and the LORD sent thunder and hail, and the fire ran along upon the ground; and the LORD rained hail upon the land of Egypt. **24** So there was hail, and fire mingled with the hail, very grievous, such as there was none like it in all the land of Egypt since it became a nation. **25** And the hails smote throughout all the land of Egypt all that was in the field, both man and beast; and the hail smote every herb of the field, and break every tree of the field. **26** Only in the land of

Goshen, where the children of Israel were, was there no hail."

Graham Philips goes on to attribute the Nile turning to blood to the eruption of Thera. Exodus 7:19 states, "And all the waters that were in the river turned to blood". Thera had corrosive toxin in its bedrock: iron oxide. Wilson points out that in the submarine eruptions that still happen at Thera, tons of iron oxide is discharged that kill fish for miles round. Over the years, scholars have attributed the plagues described in Exodus to different natural phenomena. The darkness could have been due to a particularly violent sandstorm; the hail could have resulted from freak weather conditions; and the bloodied river may have been the result of seismic activity far to the south of tropical Africa.

According to the Exodus story, the Israelites are led out of Egypt by following a "pillar of cloud by the day" and a "pillar of fire by night" (Exodus 13:21). Graham Philips attributes these phenomena to the Thera plume. The towering ash cloud over the volcano itself rose more than 48 km (30 miles). It would have been visible from the delta of Lower Egypt, given the curvature of Earth. If the Israelites had attributed the phenomena to the intervention of God, then they may well have started the exit following the direction of the sky sign, in the belief that it was a beacon leading them out of Egypt, to safety. According to Exodus 13:18: "God led the people round by the way of the wilderness towards the Red Sea". The earlier interpretation of the Hebrew words "Yam Suph" as the "Red Sea" seems to have been incorrect. In fact, according to Phillips, it could mean "sea of reeds" or "sea of seaweed", which could describe an expansive, large, shallow lake that is covered by reeds here and there.

My theory is that Lake Burullus is more likely to have been the sea of reeds in the north of Egypt; it is perfectly situated along the exodus path. Lake Burullus is separated from the Mediterranean by a narrow ridge of dry land some 50 km (31 miles) long and is finally connected to the Mediterranean through a shallow inlet. From Avaris—south Mansoura city—the Israelites could have reached west of Lake Burullus within a couple of days by following the sign in the northwestern sky. As the sky sign moved past the Israelites, they realized that they would have to travel eastward i.e. around the lake at the Rasheed village; where the French expedition to Egypt, two centuries ago, found the famous Rosetta stone. The Israelites turned to the east instead of continuing west. They walked on the Burullus ridge. They reached the shallow inlet, 1,100 meter (3,600 feet) long and 200 meter (656 feet) across and 5 meter (16 feet) deep that connects Lake Burullus to the Mediterranean Sea, and they had to cross it.

If the shallow inlet connecting the lake to the sea, usually underwater, had receded because of the tsunami force resulting from the Thera eruption, the phenomenon might indeed have saved the Israelites as the Egyptians were in close pursuit. The inlet would have turned into dry land and Lake Burullus, or the sea of reeds, would have been temporarily disconnected from the Mediterranean, as the Bible relates in Exodus 14:21 : "And the Lord caused the sea to go back by a strong east wind all that night and made the sea into dry land and the waters were divided". Exodus 14:25 tells us that when the Egyptians tried to pursue the Israelites across the sea of reeds they were hampered by the ground clogging of their chariots wheels, so they drove heavily. The tidal wave caused by Thera's eruption would have finally helped the Israelites to escape untouched. The pursuing soldiers may have attempted to follow when the tsunami tidal wave hit, washing them away. Exodus 14:29 tells us, "And the waters returned and covered the chariots, the horsemen and all the host of Pharaoh who came into the sea after them; there remains not as much as one of them". Several scholars consider it unlikely to have darkness, fiery hail, sores, a bloodied river, dead cattle and fish, swarms of locusts, flies and infestations of lice; all happening at the same time. However, I believe that there is another explanation for these events. What if the sky sign was the trace of a passing asteroid? Asteroids are sometimes called minor planets or planetoids. They are smaller than planets but larger

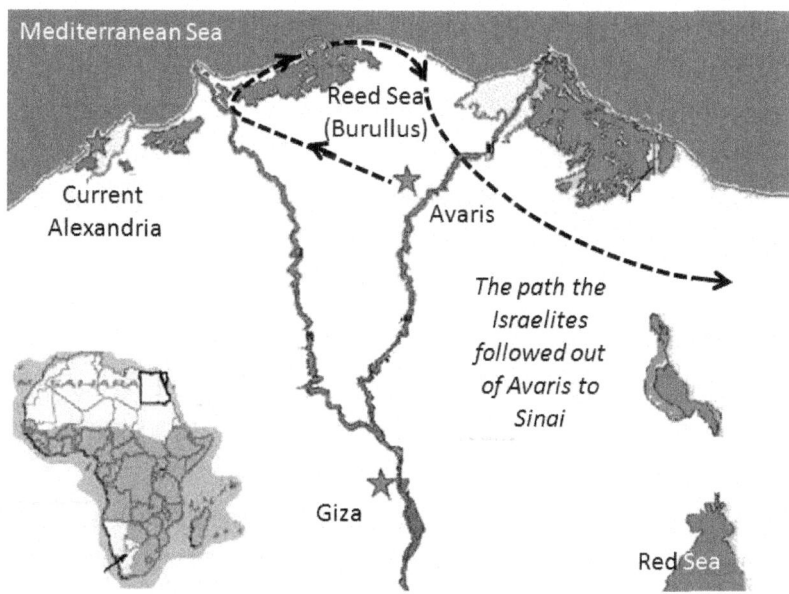

Figure 36- The migration path of the Israelites and crossing of Burullus Inlet

than meteoroids. What if it was a comet exhibiting a coma, or tail of gases and hail? This could very well fit into the Bible description of "a pillar of cloud by day" and "a pillar of fire by night". This could explain the sky sign that the Bible claims guided the Israelites out of Egypt. Not only could its magnetic pull have tilted the Inner Core magnet of the Earth, causing great Earthquakes, such as the Thera eruption, but could have also reduced the vertical component of the magnetic force and thus slowed down Earth's rotation, so that one night lasted three days long. "And Moses stretched forth his hand toward heaven; and there was a thick darkness in all the land of Egypt three days". The fiery hail covered the Egyptian sky, could have finally found its way to the Nile and turned it red. Depending on the mass of the planetoid and its speed and distance from the Earth, a negative pressure could have developed in the atmosphere at the point of proximity. A plume of air could have developed with a crest that pointed to the planetoid. Such a suction of air would have created a low atmospheric pressure at the surface of the Earth. This would have brought gale winds storming from the East to try to balance the drop in pressure "And the Lord caused the sea to go back by a strong east wind". But, where did this planetoid come from, and where did it go? Is it going to come back again? Had it visited our solar system before? When and at what frequency?

Asteroid impact at sea may be the most problematic to life on Earth. An asteroid (Hancock, 1998) with a radius of 200 m (666 ft) that drops anywhere in the mid-Atlantic will produce deepwater waves that are at least 5 m (16.6 ft) high when these waves reach American African and European coasts. As the wave approaches the shallow sea bed at the coastline, it steepens into a tsunami at 200 m (666 ft) high and hits the coast at pulse duration of two minutes between two successive waves. An asteroid that is 500 m (1,665 ft) in radius would produce a deepwater wave 50 to 100 m (333 ft) high. As the tsunami height could be amplified by a factor of 20 or more, in the run up to the continental shelf, a tidal wave of few kilometers becomes imminent. On a ship at sea, one would hardly notice the swell. But approaching the shoreline a wave slows down and increases in height as it enters shallow water. There is a piling up of water as the forward part of the wave slows down. Sky observatories are involved in a systematic search for near-Earth asteroids. It is reported by the American National

Aeronautics and Space Administration (NASA)[21] that, back in 1995, they found twenty to thirty near-Earth objects, the smallest of which is only 6 m (20 ft) across, every year. The number discovered grew drastically to have reached four hundred by end of 2008. To the extent practical, NASA, in coordination with the American Department of Defense and other countries, (Hancock, 1998) started, in the mid-1990s, to identify and catalogue the orbital characteristics of all comets and asteroids that are greater than 1 km (0.6 miles) in diameter and are in an orbit around the Sun that crosses the orbit of Earth. "During the last two centuries, astronomers have learned a great deal about the solar system and about the near-Earth space. As our planet orbits the Sun at a speed of 110,000 km/hour (66,000 mile/hour), it passes through lumpy streams of cosmic debris", said Hancock. Most of the rubble is the size of tiny meteoroids and burns in the atmosphere.

One worrying feature is that many impacts or near-impacts appear to have involved groups of projectiles rather than individual ones. Craters are evidence of impacts. However, because the surface of Earth is dynamic and is subjected to erosion and disposition, and because water covers 71 percent of the planet few craters remain today. In 1989 an asteroid with an estimated diameter of 500 m (1,665 ft) crossed Earth's path, missing it by only six hours. Had it struck Earth, it would have caused a disaster unprecedented in human history. The reason is that these projectiles carry huge reservoirs of kinetic energy. The energy of a moving body is equal to the product of half its mass and the square of its velocity. Upon impact, the kinetic energy surrenders explosively, generating terrifying shock waves. Upon impact, an asteroid will be brought to a halt in a distance equals to its own diameter. Pressures of several millions of atmospheres and shock temperatures of tens of

---

[21] http://neo.jpl.nasa.gov/stats/

thousands of degrees are immediately generated.

Emilio Spedicato, Professor of Operations Research at the University of Bergamo in Italy reports that the atmospheric disturbance resulting from collision with a 10 km (6 miles) object would be colossal and extends over hemispheric areas. It can be estimated that if 10 percent of the initial energy of the flying object goes into the blast wave, at 2,000 km (1,200 mile) from the impact point, the wind velocity would be 2,400 km/hour (1,440 mile/hour) with duration of 0.4 hours and air temperature of 480 degrees Celsius. At 10,000 km (6,000 mile), these numbers would be, respectively, 100 km/hour (60 mile/hour), 14 hours and 30 degrees. Victor Clube, dean of the Astrophysics Department of Oxford University and Bill Napier research Astronomer of the Royal Armagh Observatory have calculated that if such an impact were to occur in India, it would flatten forests in Europe, setting them ablaze. As a result of hundreds of fires, 50 million tons of smoke would be ejected upwards, rising to an altitude of 10 km (6 mile) and darkening the skies around the globe.

Abraham saw a planet. The chapter of Al Anaam Surat, verses 76-77, of the Qur'an describes Abraham's bewilderment, trying to find God:

"When the night grew pitch dark upon him he beheld a <u>planet</u>. He said: This is my Lord. But when it set, he said: I love not things that set. **76** And when he saw the Moon uprising, he exclaimed: This is my Lord. But when it set, he said: Unless my Lord guides me, I surely shall become one of the folk who are astray. **77** And when he saw the Sun uprising,

"فَلَمَّا جَنَّ عَلَيْهِ ٱلَّيْلُ رَءَا كَوْكَبًا قَالَ هَٰذَا رَبِّى فَلَمَّا أَفَلَ قَالَ لَا أُحِبُّ ٱلْأَفِلِينَ (٧٦) فَلَمَّا رَءَا ٱلْقَمَرَ بَازِغًا قَالَ هَٰذَا رَبِّى فَلَمَّا أَفَلَ قَالَ لَئِن لَّمْ يَهْدِنِى رَبِّى لَأَكُونَنَّ مِنَ ٱلْقَوْمِ ٱلضَّآلِّينَ (٧٧) فَلَمَّا رَءَا ٱلشَّمْسَ بَازِغَةً قَالَ

he cried: This is my Lord! This is greater! هَـٰذَا رَبِّى هَـٰذَآ أَكْبَرُ ۖ فَلَمَّآ أَفَلَتْ And when it set, he exclaimed: O my قَالَ يَـٰقَوْمِ إِنِّى بَرِىٓءٌ مِّمَّا people! Lo! I am free from all that ye تُشْرِكُونَ" associate (with Him)."

How did Abraham see a planet in the sky with his bare eyes? Staring at the night sky, excluding the crescent or full Moon, one would only notice many sparkling objects. Unless one has a space telescope, a device only invented at the beginning of the seventeenth century; no one could tell whether a sparkling object belongs to a star or to one of the planets of the solar system. All objects appear tiny and sparkling. They all look the same to the naked eye. What if there was a planet that was close enough to be seen as clearly. The planet would not be larger but would have only appeared as the Moon and the Sun because of its proximity to Earth. Otherwise, how did Abraham see it and thought of it as the creator of all? If the last visit of such a planet (let us call it the magneto-planet or *mPlanet* for simplicity) to the solar system occurred at the time of Moses 3,550 years ago as discussed earlier, then when was it that Abraham saw it?

Most scholars estimate the time of Abraham to be 2,000 years BC. According to the Old Testament of the Bible (Mozes, 2007) and according to (Abu-Salieh, 1999), Abraham is believed by Jews and Muslims to be the first male to be circumcised in the history of mankind. In the report of Professor James Swain's presidential address to the Bristol Medico-Chirurgical Society, the statement is made that "there are no representations of Egyptian circumcision earlier than the time of Ramses II (1310-1243 BC)". Later in the British Medical Journal of March 28th 1908 (p.732, first footnote) G. Elliot Smith, Anatomical Department, the University of Manchester, (Smith, 1910) drew the attention to a series of ancient Egyptian pictures of the operation of circumcision, which were carved 2,000 years before the time of Ramses II. He also referred to the bodies of early prehistoric men from Egypt who had been circumcised a thousand year earlier still. Therefore and since he was the first male to be circumcised, Abraham seemed to have lived some 7,000 years ago? A 3,550 years *mPlanet* cyclic trip to the solar system becomes more likely.

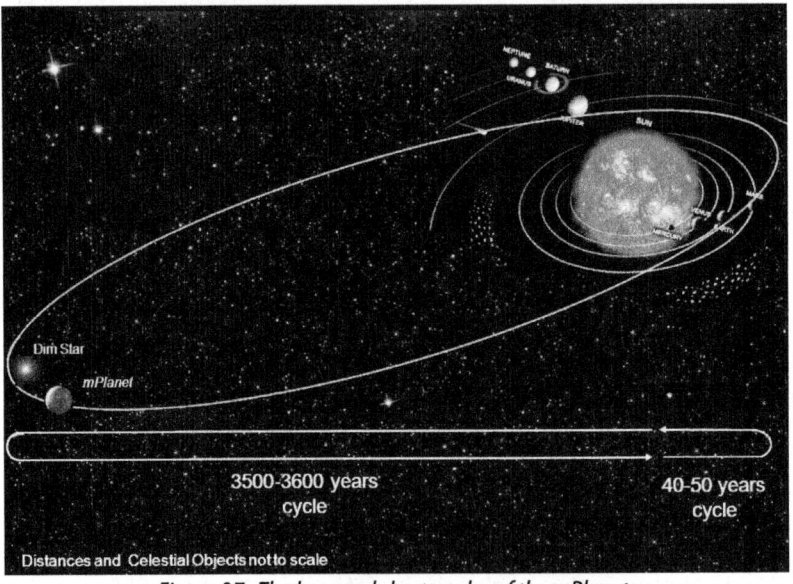

Figure 37- The long and short: cycles of the mPlanet

One of the explanations for the periodicity of the visit is that such a planet oscillates around two stars: our Sun as well as a dim star that is quite far from our solar plateau. The *mPlanet* would have to circle in an elliptic path around the two stars that are situated at the focal points of the planet's elliptic path. Not only will the *mPlanet* visit Earth in a cycle of 3,550 years; it will also travel farther our Sun for quite a mileage before it swings back, thanks to the Sun's gravitational pull. There are, therefore, two cycle times for the *mPlanet* to come close to Earth, one that takes roughly 3,500-3,600 years and another that takes between 40-50 years, as will be evidenced later in this chapter. In 1665, Sir Isaac

Newton was the first to show that the law of the force that pulls the Moon toward Earth is the same law that makes an apple fall. Newton concluded that not only does the Earth attract both an apple and the Moon, but everybody in universe attracts every other body. This tendency of bodies to move towards each other is called gravitation. Newton proposed the law, which is now known as Newton's Law of Gravitation. "The force that pulls two bodies to one another is proportionate to the product of their masses and is inversely proportionate to the square distance separating them". So the closer the bodies approach one another, the fiercer the gravitational force becomes. Such a force, when exerted on either body, will help it accelerate. Acceleration is inversely proportionate to the mass of the body. The *mPlanet* may have a much smaller mass than the Sun does. While both interstellar objects are exposed to same gravitational force, resulting in their coming closer to one another, the acceleration of the *mPlanet*, because it has a much smaller mass than the Sun, becomes high. This helps the *mPlanet* gain a huge amount of kinetic energy that gets even higher when the *mPlanet* revolves around the Sun prior to bouncing back in its 3,550-year trajectory path before a newer rendezvous with the solar system. Apollo 13 was the third manned mission by NASA, intended to land on the Moon[22], but a mid-mission technical malfunction forced the lunar landing to be aborted. The mission was launched on April 11, 1970. Two days later, while the mission was en route to the Moon, a fault in the electrical system of one of the service module's oxygen tanks produced an explosion that caused both oxygen tanks to fail and also led to a loss of electrical power. The command module remained functional on its own batteries and oxygen tank, which were designed to support the vehicle only during the last hours of flight. The main question was, 'How to get back

---

[22] http://en.wikipedia.org/wiki/Apollo_13

safely to Earth?' The malfunction caused the abortion of the mission. It fortunately happened on the way to the Moon, when the lunar module was still available with its full complement of fuel and food. Had the malfunction occurred after the landing or on the return to Earth after the lunar module had been jettisoned, the crew would not have survived. The crew decided to move to the lunar module. The chances that the boost would be sufficient to put them on a correct trajectory to Earth were not high. Instead of propelling it towards Earth, the crew directed the lunar module to revolve around the Moon, knowing that the gravitational force that would be established between the Moon and the lunar module would be great enough to help accelerate the tiny lunar module mass straight to Earth. Although not detected yet by any space observatory, I believe that the *mPlanet* went through same process at the other side of its elliptic path, and the dim star had helped accelerate it back to our solar plateau. Coming close to the Sun, increases the chances of magnetic pull between the two bodies. The surface of the Sun consists of hydrogen (about 74 percent of its mass, or 92 percent of its volume), helium (about 24 percent of mass, 7 percent of volume), and trace quantities of other elements. The Sun has "quakes" that are similar to Earthquakes. Seismic waves produced by Sunquakes can shake the Sun to its very center, just as Earthquakes can cause the entire Earth to vibrate. However, Sunquakes involve much more energy than their terrestrial counterparts. An observed Sunquake, which was produced by a perfectly ordinary solar flare, was equivalent to an Earthquake of a magnitude 11.3 on the Richter scale. That is 40,000 times more energy than the devastating San Francisco Earthquake of 1906, at a magnitude of 8 on the same scale[23]. The Sunquakes will generate immense and multiple solar flares. Solar flares affect all layers of the solar atmosphere—photosphere, corona, and

---

[23] http://en.wikipedia.org/wiki/Quake_(natural_phenomenon)#Sunquake

chromosphere—heating plasma to tens of millions of kelvins and accelerating electrons, protons, and heavier ions to near the speed of light. They produce radiation across the electromagnetic spectrum at all wavelengths, from radio waves to gamma rays. The sudden release of magnetic energy stored in the corona powers the flares. The frequency of occurrence of solar flares varies, from several per day when the Sun is particularly active to less than one each week when the Sun is quiet. Large flares are less frequent than smaller ones. Solar activity varies within an eleven-year cycle known as the solar cycle. At the peak of this cycle there are typically more Sunspots on the Sun, and hence more solar flares. X-rays and ultraviolet radiation emitted by solar flares can affect Earth's ionosphere and disrupt long-range radio communications. Direct radio emission at deci-metric wavelengths may disturb the operation of radar and other devices operating at these frequencies. The arrival to Earth of protons and positrons in plasma state amongst other particles will have an impact on ferromagnets and electromagnets, as we shall see later in chapter four, and therefore disable the function of most electromechanical machinery.

When Earth stopped: The Bible tells us that Moses and the Israelites had spent forty years in the wilderness of the Sinai Peninsula. Upon Moses death, at the close of the forty years, Joshua, son of Nun, assumed the leadership and marched against the Hittites and others. In Joshua-1:1-5 the Lord passes his commands:

The LORD Commands Joshua
"**1** After the death of Moses the servant of the LORD, the LORD said to Joshua son of Nun, Moses' aide: **2** "Moses my servant is dead. Now then, you and all these people, get ready to cross the Jordan River into the land I am about to give to them—

to the Israelites. **3** I will give you every place where you set your foot, as I promised Moses. **4** Your territory will extend from the desert to Lebanon, and from the great river, the Euphrates—all the Hittite country—to the Great Sea on the west. **5** No one will be able to stand up against you all the days of your life. As I was with Moses, so I will be with you; I will never leave you nor forsake you."

The Bible goes on to describe the march of Joshua's army against the coalition of the kings of the Hittites, Amorites, Canaanites, Perizzites, Hivites and Jebusites after having captured Jerico and Ai. It was at this time, between forty to fifty years after the events described in Exodus, when the *mPlanet* swayed back and came closer to the Earth for the second time in less than fifty years. The Bible tells us of the day the Sun stopped. It seems more likely that the Earth stopped temporarily than that the Sun did thanks to the magnetic and gravitational pull of the *mPlanet*. Here is what the Bible Joshua 10:1-15 tells us:

The Sun Stands Still

"**11** As they fled before Israel on the road down from Beth Horon to Azekah, the LORD hurled large hailstones down on them from the sky, and more of them died from the hailstones than were killed by the swords of the Israelites. **12** On the day the LORD gave the Amorites over to Israel, Joshua said to the LORD in the presence of Israel: "O Sun, stand still over Gibeon, O Moon, over the Valley of Aijalon". **13 So the Sun stood still** and the Moon stopped, till the nation avenged itself on its enemies, as it is written in the Book of Jashar. The

Sun stopped in the middle of the sky and delayed going down about a full day. **14** There has never been a day like it before or since, a day when the LORD listened to a man. Surely the LORD was fighting for Israel! **15** Then Joshua returned with all Israel to the camp at Gilgal."

In other words, the day lasted longer. The hour became longer or **MORE THAN 60 MINUTES** as this book is titled.

What about Noah, who is estimated to have lived between 8,000 and 9,000 BC? Could it be that 10,600 years ago, the *mPlanet* had approached our solar system? If so, would the same pattern of Earth Changes have been repeated? Here is what would probably have happened, given the historic citing of the holy books; the magnetic pull of the *mPlanet* would have caused the core of the Earth to tilt, and the magnetic field to tilt along. As a result of the magnetic pull, it would have protruded into space, and so it appeared to get weaker. The tilting of the magnetic core would have been manifested by the magnetic poles wandering away from their original positions. The ice caps formed over thousands of years, would have become unshielded and would have suddenly melted in a way similar to what we have started to experience recently. The grand finale would have taken the world by surprise, as the final tilt left the remaining ice unprotected, causing a sudden fragmentation of giant ice blocks to fall into the ocean, producing tidal waves and tsunami along the shores. The ice meltdown will elevate the sea water to an unprecedented level. Let us consider it one step at a time:

1) The magnetic pull of the *mPlanet*, at a distance, started weak and frail and got stronger as it approached the solar system.

2) The Magnetic Field Force Lines (MFFL) circumventing the Earth got less and less as some of it bonded, through space, with the MFFL of the *mPlanet*.
3) The weaker the MFFL surrounding the Earth became, the more space is given to the Sun charged particles to increase its speed and spiral-radius motion between the Earth magnetic poles.
4) The faster, longer spiral-radius motion the protons pick up around the magnetic field force lines, the more thermal energy is produced as they collide with air molecules of the Thermosphere layer of the atmosphere. A thermal energy map for the Thermosphere is emerged with hottest spot high above the magnetic equator where the MFFL are weakest due to the curvature of our planet.
5) The more energy radiated in the Thermosphere layer, the more thermal energy reaching the surface of the Earth and a Global Warming is observed. Fire sparked easily in dry wood and surface heat became unbearable.
6) The Inner Core MFFL that bonded with that of the *mPlanet* started to cause the Inner Core to tilt, producing tremors and quakes.
7) The magnetic poles followed every new position that the Inner Core assumed and accordingly started to wander away from original locations. The whole MFFL started to tilt.
8) Following the Thermosphere thermal map, the temperature belts of the Earth are centered on the wandering magnetic pole. They, too, started to tilt resulting in Climate Change.
9) The ice caps started to melt down and initially eroded from the side that the high intensity MFFL was tilting away from.
10) The water level in the ocean started to rise.
11) The vaporization process took care of the rising ocean level.
12) More clouds were formed and excess water vapour became airborne, leading to gale winds and rain.

13) Increased precipitation resulted all over the planet, and mostly along the magnetic equator, where evaporation is highest.
14) Some parts of the Earth became cooler, and other parts became warmer. There was more rain in parts, and there was drought in others.
15) At the interim stage, rain kept pouring and floods exaggerated. Tremors proliferated as the Inner Core tilting speed increased.
16) At the climax stage, the tilting was so rapid that the whole ice caps at the North and South Poles melted in hours, not days, weeks, or months.
17) The sudden discharge of ice chunks from the Antarctic into the ocean produced huge waves, thousands of kilometers away. The tsunami earned few hundred meters height as the tidal wave reached shallow coastal lines.
18) The tilting of the Inner Core produced a different gap angle between the Earth magnetic axis and rotation axis, leading Earth spin speed to change. The more spin speed, the more bulged oceans will accumulate around the equatorial belt and a great flood is born.

I was always wondering, how come the human maximum life span at the time of Noah was close to a thousand years of age; whereas, it is currently decreased to a hundred years? The answer could be simple when we reflect on what was stated earlier in chapter one; "The pineal gland is activated by light, and it controls the various bio- rhythms of the body. It works in harmony with the Hypothalamus gland, which directs body's thirst, hunger, sexual desire and the biological clock that determines our aging process". Chapter two tells us; "Lorentz forces are created of pairs of equal magnitude and opposite directions along the perimeter of the bulged Outer Core. The perimeter torque of such pairs of forces keeps the spinning of the planet going counterclockwise for the moment" and "The Lorentz force is proportionate to the strengths of the electric current, the magnetic force and the tilt angle between the magnetic field and the spinning axis". If the tilt angle becomes larger and the magnetic poles will settle finally in India and Brazil for example, the Earth will slow down its spinning speed about its axis and the length of day and night will, therefore, get longer. Earth would still complete a full circle around the Sun in the same time it experiences nowadays, in what we call a solar year. However, the succession of day and night will be less frequent than 365 per solar year. If the succession of day and night, during a year for example, is brought down by 10 fold i.e. to 36, the pineal gland aging process will slow down so that the life span, in terms of solar years, will increase in the same proportion i.e. 10 fold. The maximum life span could jump from 90 to 900 years. Could that sequence of events repeat itself, now that we are approaching 3,550 years from the last time the *mPlanet* paid the solar system a visit? Not only that, but would the Moon split in two as referenced in the Surat Al Qamar of the Qur'an?

"The hour drew nigh and the Moon was rent in twain. (1)"   "اقْتَرَبَتِ ٱلسَّاعَةُ وَٱنشَقَّ ٱلْقَمَرُ (١)"

The Moon's a balloon, is a 1972 bestseller authored and later played as a movie by David Niven. Neither the book nor the movie was science-based or related to the Moon. But, could the Moon be hollow like a balloon? In 1997, Jim Marrs discussed in his book *Alien Agenda* (Marrs, 1997) what could be considered the greatest unidentified flying object orbiting Earth. During the Apollo missions, seismographic equipment was placed at six separate sites on the Moon. Up to three thousand Moonquakes were detected during each year between 1969 and 1977, when the equipment ceased to operate. Most of the vibrations were quite small and were caused by meteorite strikes or falling booster rockets. There are many indications that the Moon could be hollow. The mean density of the Moon is about 3.34 grams per cubic centimeter, which is significantly different from the 5.5 grams per cubic centimeter of the Earth's Mantle. Nobel chemist Dr Harold Urey has suggested that the density question may be answered by what he termed "negative mascons" or large volumes inside the Moon where there is either matter much less dense than the rest of the Moon or simply a cavity. Massachusetts Institute of Technology (MIT)'s Dr. Sean C. Solomon, Director at the Department of Terrestrial Magnetism, Carnegie Institution of Washington wrote, "The Lunar Orbiter experiments vastly improved our knowledge of the Moon's gravitational field, indicating a possibility that the Moon might be hollow". A natural planet cannot be a hollow object! The most startling evidence that the Moon could be hollow came in 1969 when the Apollo 12 crew, after returning to the command ship, sent the lunar module back crashing onto the Moon, thus creating an artificial Moonquake. The lunar module struck the surface about 64 km (40 miles) from the Apollo landing post, where ultrasensitive seismic equipment recorded something both unexpected and astounding: The Moon reverberated much like a bell for more than an hour. The vibration wave took almost eight minutes to reach a peak, and then decreased in intensity. The phenomenon was repeated when Apollo 13's third stage was sent crashing onto the Moon by radio command, striking with the equivalent of eleven tons of TNT. According to NASA (Marrs, 1997), this time the Moon "reacted like a gong". Reverberations lasted for three hours and twenty minutes. It seems apparent that the Moon has a tough, hard outer shell and a light or nonexistent interior. The Moon's shell contains dark minerals such as titanium, used on Earth in the construction of aircraft and space vehicles. Experts were surprised to find lunar rocks bearing brass, mica and amphibole in addition to near-pure titanium. "They conclude that it is the large amount of titanium in the black mineral illuminate that gives the dark tone to the lunar seas". Uranium 236 and neptunium 237, elements not previously found in nature, were discovered on Moon rocks, according to the Argon National Laboratory", said Marrs. While still trying to explain the presence of these materials, scientists were further startled to learn of the rust-proof iron particles in a soil sample from the crater Mare Crisium. In 1976, the Associated Press reported that the Soviets had announced the discovery of iron particles that do not rust in samples brought back by an unmanned Moon mission in 1970. Iron that does not rust is unknown on Earth and well beyond present Earth technology.

The Moon splits was also stated (Hancock, 1998) as Graham Hancock shares with us what was documented by a twelfth-century monk, Gervase of Canterbury, whose chronicle is highly regarded as a work of history. He wrote that on the evening of June 25, 1178, five friends were sitting out after dark on the outskirts of the English cathedral city of Canterbury, chatting. The sky was cloudless and a bright new Moon was rising, with its horns tilted towards the east. Then, suddenly:

> "The upper horn split in two. From the midpoint of the division a flaming torch sprang up, spewing out, over a considerable distance, fire, hot coals and sparks. Meanwhile, the body of the Moon, which was below, writhed as if it were in anxiety. The Moon throbbed like a wounded snake. Afterwards it resumed its proper state".

The only reasonable explanation is that if the Moon is formed of two joined halves, it will keep so, according to Newton's first law of motion (a body at rest stays at rest, and a body in motion stays in motion, unless it is acted upon by an external force), until a passing meteoroid is either fast enough or massive enough or both to pull one half of the Moon apart while the other half remains highly influenced by the Earth's pull. There will have to be an equilibrium point at which the gravitational forces among the four bodies come to balance with one another. On top, the centrifugal force pushing revolving objects away, such as the Moon halves, will ensure that the distance is maintained among the four bodies. This helps to prevent head-to-head crash.

"The Hour" is mentioned in the holy book of Qur'an. The "End of Days" is similarly mentioned in the Bible. The term refers to the time when severe turmoil takes people by surprise. Could it be that because when Earth's spin slows down, the hour that we know to be sixty minutes will get longer? It currently takes twenty-four hours for the Earth to complete a whole revolution about its axis. Could it, at one point in time, take forty or sixty hours? The Earth could eventually cease to revolve or perhaps spins in the reverse direction. People will watch the Sun rising from the west. It might be exciting to some, but the scale of volcanoes, earthquakes, ashes in the air, gigantic tidal waves coming from ocean bed movements and ice caps sudden melting, will not let anyone to enjoy such a scene.

Remarkably, Surat Al Qamar not only introduces a phenomenal event, the timing of which is consistent with astronomical events, but it speaks of the prophets in whose times the cycle of the *m-planet* approaching the solar system had repeated itself. These were i) *Noah*, ii) the tribe of Lot, nephew of *Abraham*, and iii) the Pharaoh who reigned at the time of Moses. Each is respectively separated by 3,550 years.

"The folk of Noah denied before them, yea, they denied Our slave and said: A madman; and he was repulsed. 9 So he cried unto his Lord, saying: I am vanquished, so give help. 10 Then We opened the gates of heaven with pouring water 11 And caused the Earth to gush forth springs, so that the waters met for a predestined purpose. 12"

"كَذَّبَتْ قَبْلَهُمْ قَوْمُ نُوحٍ فَكَذَّبُوا عَبْدَنَا وَقَالُوا مَجْنُونٌ وَازْدُجِرَ (٩) فَدَعَا رَبَّهُ أَنِّى مَغْلُوبٌ فَانْتَصِرْ (١٠) فَفَتَحْنَا أَبْوَابَ السَّمَاءِ بِمَاءٍ مُنْهَمِرٍ (١١) وَفَجَّرْنَا الْأَرْضَ عُيُونًا فَالْتَقَى الْمَاءُ عَلَى أَمْرٍ قَدْ قُدِرَ (١٢)"

"The folk of Lot rejected warnings. 33 Lo! We sent a storm of stones upon them (all) save the family of Lot, whom We rescued in the last watch of the night, 34 As grace from Us. Thus We reward him who giveth thanks. 35"

"كَذَّبَتْ قَوْمُ لُوطٍ بِالنُّذُرِ (٣٣) إِنَّا أَرْسَلْنَا عَلَيْهِمْ حَاصِبًا إِلَّا آلَ لُوطٍ نَجَّيْنَاهُمْ بِسَحَرٍ (٣٤) نِعْمَةً مِنْ عِنْدِنَا كَذَلِكَ نَجْزِى مَنْ شَكَرَ (٣٥)"

"And warnings came in truth unto the house of Pharaoh 41 Who denied all Our revelations. Therefore We grasped them with the grasp of the Mighty, the Powerful. 42 Are your disbelievers better than those, or have ye some immunity in the scriptures? 43 Or say they: We are a host victorious? 44 The hosts will all be routed and will turn and flee. 45 Nay, but the Hour their appointed tryst, and the Hour is more wretched and bitterer. 46"

"وَلَقَدْ جَاءَ آلَ فِرْعَوْنَ النُّذُرُ (٤١) كَذَّبُوا بِآيَاتِنَا كُلِّهَا فَأَخَذْنَاهُمْ أَخْذَ عَزِيزٍ مُقْتَدِرٍ (٤٢) أَكُفَّارُكُمْ خَيْرٌ مِنْ أُولَئِكُمْ أَمْ لَكُمْ بَرَاءَةٌ فِى الزُّبُرِ (٤٣) أَمْ يَقُولُونَ نَحْنُ جَمِيعٌ مُنْتَصِرٌ (٤٤) سَيُهْزَمُ الْجَمْعُ وَيُوَلُّونَ الدُّبُرَ (٤٥) بَلِ السَّاعَةُ مَوْعِدُهُمْ وَالسَّاعَةُ أَدْهَى وَأَمَرُّ (٤٦)"

The Mayan calendar is defined as a long time cycle of 1,872,000 days (Gilbert, 2007). It occurs between two periods where considerable changes happen to Earth. By dividing by 365.25, historians translated such a number of days into 5,125 years. I believe that such as resultant number of solar years is incorrect. The Aztecs believed that a new cycle could be born on a 52-year boundary count. They used to run a grand festival every 52 years in anticipation of the birth of a new Sun. That can only be realized, if Earth stops spinning about its axis for some time and the night grew longer than normal for a couple of days. As the Earth starts to spin again, the Sun (called a new Sun by the Aztecs and the Mayans) reemerges in the sky and assumes its normal course. Therefore, it is expected that the Mayan long cycle be divisible by 52 without a remainder. When we divide 5,125 year by 52 years, we get 98.56 (i.e. there is a fraction remainder!). Joseph Goodman who first deciphered the Mayan Calendar must have overlooked the Aztec record of 52-year boundary count.

Here I recall what I explained earlier and that there were times or cycles where the rotation of the planet about its axis was slower than our current time. What if at the time of the great eruption of Thera (Santorini) and the Exodus of the sons of Israel from Egypt, took place exactly 1,545 BC, that the short cycle of Earth changes is 3,562 years, and that the Mayan cycle calendar is twice as long? Here we can divide the Mayan long cycle days into two periods: 1,301,020 days of 365.25 days per year, equivalent of 3,562 years and 570,979 days over 3,562 years resulting in 160.30 days per year, which means a 13- day month; a number the Mayan used to cherish and live by. Upon dividing Mayan long cycle of 7,124 years by 52, we get 137 (an integer number without a remainder) which conforms in with what the Aztec believed, that a new cycle could be born on a 52 year boundary interval. In other words the Mayan long cycle started 5,107 BC and could complete in 2017 AD.

This means that no Earth changes to mention will occur on December 21st 2012 AD. If the Mayans are right, a cycle ends and a new one begins in 2017 AD.

Other observations from earlier civilizations; The Avestic Aryans of Iran who lived from 8,000 BC, spoke of three epochs (Tilak, 1903) of creation prior to our current one. The first of which witnessed sudden and drastic climate change that turned the climate in which they lived from 5 months of winter/ 7 months of summer to 10 months of winter/ 2 months of summer and that length of day and night was long. If we assume that the Avestic Aryans had lived in the centre land of current Iran, this meant that before the commencement of the first epoch a magnetic pole, where an Ice Cap resided, must have been located within some 3,600 km (2,250 mile) distance (Continental Belt) from the centre of Iran. Given that some 10,000 years ago there was no trace of Reversed Magnetic Field (Plasmoids) in Antarctica, the planet must have been spinning very slowly about its axis to the extent that Inner Core emitted electrons flew in straight lines in the molten Outer Core and therefore no induced magnetic force was generated. For the planet to spin slowly, the gap angle between Earth magnetic axis and the axis of rotation must have been close to 90°. Plotting the distance and the gap angle, means that such a magnetic pole must have resided somewhere in India main land or Indian Ocean at 5°N latitude. Applying the same logic after the commencement of the first epoch, we can confirm that the magnetic pole, where a new Ice Cap formed, must have been relocated within a Cold Polar Belt of some 2,600 km (1,625 mile) distance from the centre of Iran; that is somewhere in India at 15°N latitude.

Since the two Magnetic Poles are not aligned opposite to each other on the surface of Earth and since each pole moves at a different rate and direction, it is not true that if one magnetic pole existed in India the other pole will be located in the Pacific Ocean west to Ecuador. Confusion exists over why the ice in Antarctica dates back, apparently for tens of thousands of years. Unlike the rings in a tree, which show its age, layers of ice do not show what is absent. Polar ice reflects only that portion of ice, which has not melted away. Many shifts are slight, thus causing partial melting or melting on only one side, If one analyses the last few shifts, it becomes apparent that the north magnetic pole in the southern hemisphere has moved slightly over Antarctica and nearby ocean.   When an ice cap forms over water near land, the land mass retains ice under the influence of such large block of floating ice cap.

The Toba Indians (Hancock, Fingerprits of the Gods, 1995) of the Gran Chaco region that sprawls across modern borders of Paraguay, Argentina and Chile, in South America at 21°S latitude still repeat an ancient myth of a great cold period that was accompanied by a great darkness. As we reflect on the great cold over central Latin America, while the Antarctic ice kept intact some 10,000 years ago, or in other words the change from Continental Belt to Cold Polar Belt or from 3,600 km (2,250 mile) to 2,600 km (1,625 mile) closer to the magnetic pole, we find a possible location for the north magnetic pole to have been at 70.00°W longitude and 60.00°S latitude. But, what caused such a movement of Inner Core magnet, as manifested by the relocation of its magnetic poles on the surface of the Earth? Was it the same *mPlanet* that was observed in the following epoch by Abraham? The Cahto Indians of California tell that amidst a great darkness, the Sun, instead of holding its course across the sky, seemed to speed crookedly overhead and to rush down in wrath like a meteor. The grass withered; the crops were scorched; the woods went up in fire and smoke. I believe that Earth spin about its axis had ceased and that the night grew longer over the western hemisphere as a consequence of the same *mPlanet*. The Cahto Indians had mistaken the intruding planet for the Sun and they saw it moving as a meteor. I also believe that the magnetic pull of the inner core of such a planet has weakened the Earth magnetic field by directly pulling some of its force lines so that the magnitude and intensity of the total magnetic force lines circumventing the surface of the Earth became less. A weaker, less numbered magnetic field force lines results in the capture of Sun charged protons to gain higher speed and experience longer spiral-radius motions between the two magnetic poles of our planet. The increased collisions of sped, longer spiral-radius moving protons with Thermosphere air molecules generate higher thermal energy, which at a point in time became scorching enough to spark fire at dry woods and forests.

Cycle features vary greatly. Whereas the short cycle is 3,562 years, we do not know what is the number of days per solar year? What is the spin speed of Earth about its axis? What is the location of the magnetic pole, the Temperature Belts and the resultant Climates? Some 10,600 years ago there were no traces of Plasmoids (Reversed Magnetic Field (RMF) at the Antarctica. This means that the spin speed was so slow so that the electrons flowing in the outer core did not move in spiral paths and accordingly resulted in no induced magnetic field force. As we read about the age of humans living in those ages, we find it to be in the range of a thousand year.

Since the Pineal Gland controls the aging of the human body and since it is activated by days and not by solar years, it seems that Earth had rotated slower in earlier cycles. Therefore, human life time in days is almost fixed regardless of the cycle features according to the following formula:

The Cycle

> Human Days/Life = (Number of Days/Year)$_i$ X (Number of Years/Life)$_i$
> Where $_i$ denotes the cycle number.

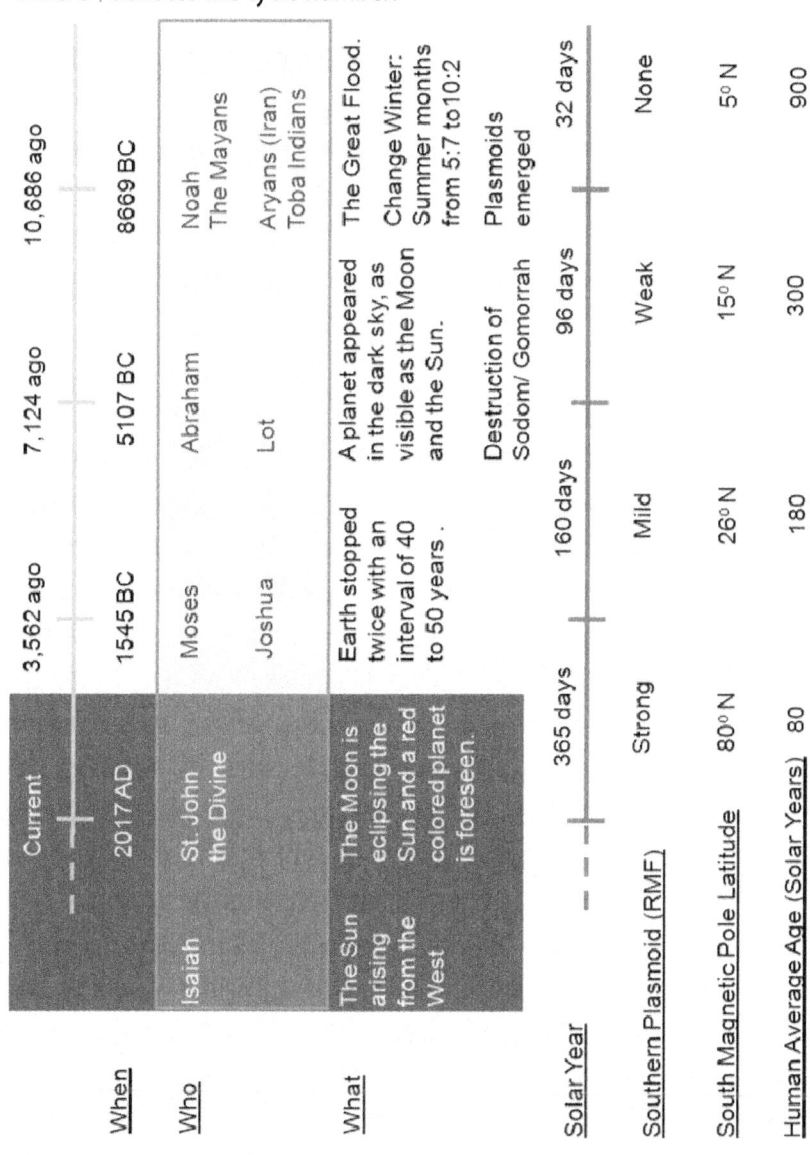

When	Current	3,562 ago	7,124 ago	10,686 ago
	2017 AD	1545 BC	5107 BC	8669 BC
Who	Isaiah	Moses	Abraham	Noah / The Mayans
	St. John the Divine	Joshua	Lot	Aryans (Iran) / Toba Indians
What	The Sun arising from the West	Earth stopped twice with an interval of 40 to 50 years.	A planet appeared in the dark sky, as visible as the Moon and the Sun.	The Great Flood. Change Winter: Summer months from 5:7 to 10:2
	The Moon is eclipsing the Sun and a red colored planet is foreseen.		Destruction of Sodom/ Gomorrah	Plasmoids emerged
Solar Year	365 days	160 days	96 days	32 days
Southern Plasmoid (RMF)	Strong	Mild	Weak	None
South Magnetic Pole Latitude	80° N	26° N	15° N	5° N
Human Average Age (Solar Years)	80	180	300	900

Figure 38- Cycle timeline of Earth changes and count of days per year

In our times the average human age is 80 solar years and each year is 365 days. Some 10,600 years ago, during Noah's time, the average human age was 900 solar years as mentioned in the Old Testament which means that each year was 32 days long, as we apply the Human Days/Life formula above. Upon applying the fixed ratio of cosine the Gap Angle between the magnetic force lines and the axis of rotation to the Human Average Age at current and 10,600 years ago epochs, we find that the south magnetic pole was situated at 5°N latitude as manifested in the *Other observations from earlier civilizations* section of this chapter.

The twelfth planet: Today, scientists confirm eleven planets in the solar system including three dwarf planets. International Astronomical Union (IAU) defines a planet as "a celestial body that i) is in orbit around the Sun, ii) has sufficient mass for its self-gravity to overcome rigid body forces so that it assumes a hydrostatic equilibrium, nearly round shape and iii) has cleared the neighborhood around its orbit". Dwarf planets such as Pluto, Eris or Makemake have not cleared the neighborhood around their orbits. Could the *mPlanet* be the twelfth planet? As the twelfth planet gets much closer, its magnetic pull becomes fiercer and the tilting of the Earth's core accelerates, causing serious Earthquakes, volcanoes and Crust turmoil. What if the Dead Sea Transform fault widens between the Red Sea and the Dead Sea? Will the next cycle bring the two seas—the Red and the Dead—together? Not only will the geography change, but so will the meteorological map of the Earth. The Köppen Climate Classification System is the most widely used for classifying the world's climates. Most classification systems used today are based on the one introduced in 1900 by the Russian-German climatologist Wladimir Köppen. Köppen divided the Earth's surface into climatic regions that generally coincided with world patterns of vegetation and soils. The Köppen system recognizes five major climate types based on the annual and monthly averages of temperature and precipitation as shown in Figure 39.

1. Tropical Belt, where moist tropical climates are known for their high temperatures all year round and for their large amount of rain.
2. Dry Climates Belt, which are characterized by little rain and a huge daily temperature range and are either semiarid, steppe, or arid and desert.
3. Middle Latitude Belt, where humid middle-latitude climates' land/water differences play a large part. These climates have warm, dry summers and cool, wet winters.
4. Continental Belt, where moderate climates can be found in the interior regions of large land masses. Total precipitation is not very high, and seasonal temperatures vary widely.

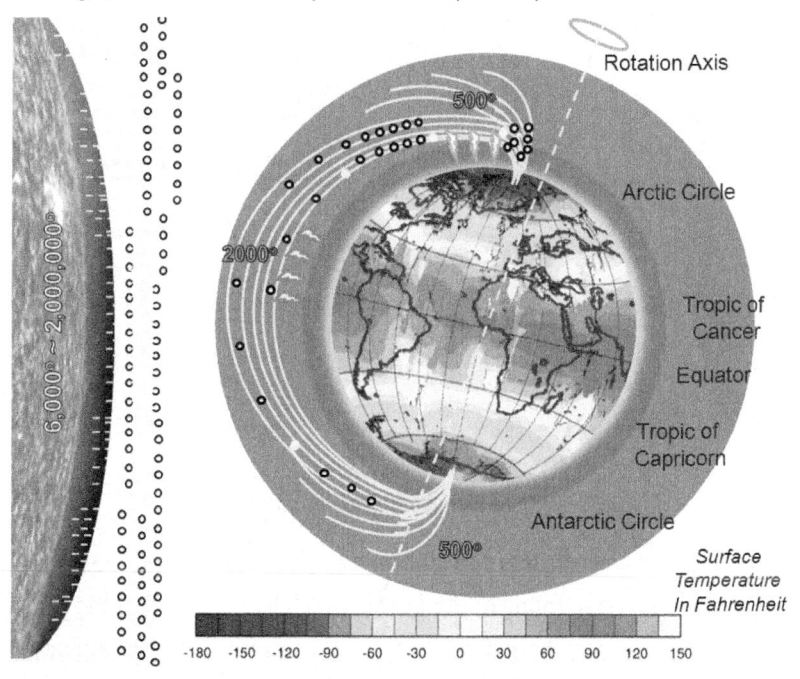

Figure 39- Temperature Belts before the poles shift

5. Cold Polar Belt, consisting of cold climates. These climates are part of areas where permanent ice and tundra are always present. The temperature rises above freezing for only about four months a year.

What if, as a result of Earth's magnetic core tilting while the Crust keeps its current geology almost intact, the magnetic poles settle halfway down the path of half-circle tilt? What will become of the magnetic shield? Will it be able to withstand the escalation of the Sun flares and accelerating electrons, protons, and heavier ions that are bombarding the Earth at near the speed of light? Birds' migration paths will obviously have to align with the tilted temperature belts.

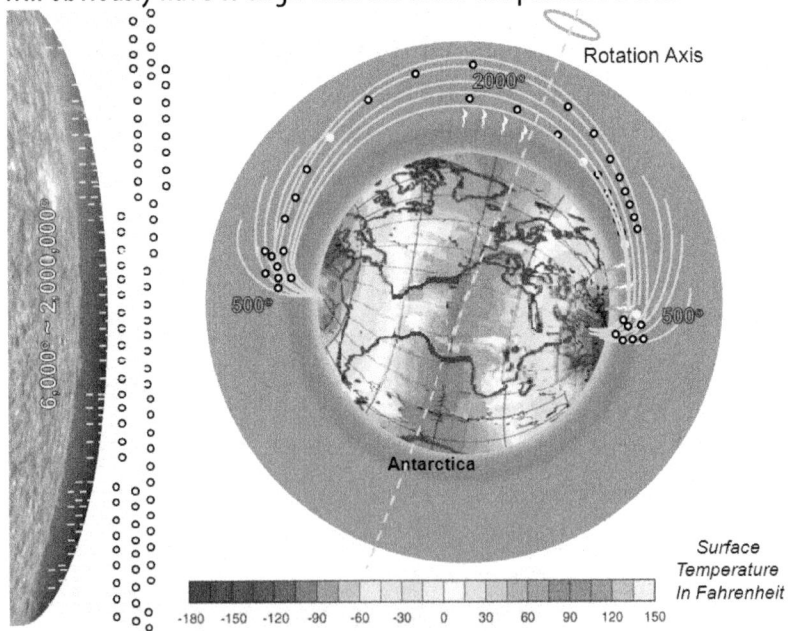

Figure 40- Potential Temperature Belts after the poles settle

A totally new agriculture map will then evolve. Mexico and the Arabian Peninsula will have many green forests much like central Europe today, whereas Somalia will enjoy a climate close to that of southern Europe. In contrast, the UK and Western Europe may experience climates of higher temperature and less precipitation as shown in Figure 40. As Earth slows down its spin, the bulged oceans will get defused between the two Tropics. The excess water will suddenly cause the sea level to rise along the shores of northern and southern hemispheres. Some may think that due to Earth slow spin for a considerable period of time that longer nights in the region not facing the Sun will be freezing cold. They would be right if the warmth reaching the surface of the Earth is sourced from the Sun's radiation.

As I manifested earlier, the Sun radiation hardly contributes to the warmth of the surface of the planet but it is the Thermosphere that does it. Figure 40 illustrates that the Thermosphere temperature distribution would keep both halves of the planet at similar but tilted Temperature Belts of current days. "The correction of the order of all things by God is as crisp and final" as it was when St. John wrote it in exile on Patmos island, where the cruel emperor Domitian banished him about sixty-four years after Jesus Christ's ascension, according to the Bible Revelation 6:12-17

> "**12** And I saw, when he had opened the sixth seal: and behold there was a great Earthquake. And the Sun became black as sackcloth of hair: and the whole Moon became as blood. **13** And the stars from heaven fell upon the Earth, as the fig tree casteth its green figs when it is shaken by a great wind. **14** And the heaven departed as a book folded up. And every mountain and the islands were moved out of their places. **15** And the kings of the Earth and the princes and tribunes and the rich and the strong and every bondman and every freeman hid themselves in

the dens and in the rocks of mountains: **16** And they say to the mountains and the rocks: Fall upon us and hide us from the face of him that sitteth upon the throne and from the wrath of the Lamb. **17** For the great day of their wrath is come. And who shall be able to stand? "

Could it be that what St. John saw was an eclipse during which the Moon obscured the Sun at the time when the *mPlanet* or twelfth planet rather is approaching and is visible, similar to the Moon in appearance but red in colour? In 2017, there is expected to be a solar eclipse over the United States on August 21st. Could the end of a cycle and the beginning of a new cycle, of changed Earth, be on August 21st 2017?

Isaiah was a prophet of God, who lived from about 740 to 681 BC. Isaiah is also the name of a book in the Old Testament of the Bible, which tells the story of Israel in his time. Isaiah predicted in this book, the birth of Jesus, hundreds of years before. In the book of Isaiah 45:5-6, it is written;

"**5** I am the LORD, and there is none else, there is no God beside me: I girded thee, though thou hast not known me. **6** That they may know from the arising of the sun, and from the west, that there is none beside me. I am the LORD, and there is none else".

Having gone through Chapter 2 of this book, we now learn how the Earth could have its spin about its axis reversed and the Sun to rise from the west.

The Mandate

Capital markets around the world started to decline in early 2007. In a two-year period, most companies enlisted in the capital market have lost over two-thirds of their market value. Many companies have inflated their assets by revaluating, on paper, their belongings according to inflated market prices. That seemed acceptable so long as the prices in the markets were on a rising trend. Comes the time when demand is growing less and prices are falling, a severe devaluation of the assets becomes a must. Hedging and fixing future buys at agreed-upon current prices worked well so long as the economy was moving on a positive trend. Fixing future buys may still be lucrative only when the bottom is reached and the economy starts to recover—but, unfortunately, not otherwise. Traders committed to future buys at the cost of prevailing prices faced a great shock when the market prices fell; they still have to meet their purchase commitments at preset high cost and resell at new lower market prices so that they can stay competitive and do not lose a market share.

It took Japan eight years from the peak to get to zero-interest rates, twelve years to reach "quantitative easing", in which the Central Bank flooded the financial system with money, and fifteen years for equities to hit bottom. In the US, much like the rest of the world, the first two milestones were reached in eighteen months, and the third one will probably be reached faster as well. The credit crisis induced a vicious cutback in spending by consumers in 2008. This, in turn, forced producers to slash production, jobs and capital spending. Governments are now trying to induce consumers to slow their desire to save by reducing borrowing costs and taxes, and improving the flow of credit. The history of recessions shows that employment is usually deteriorating when the markets hit bottom, and it continues to deteriorate during the market recovery. This phenomenon reflects the restoration of profit margins and the right-sizing of labor costs, as well as inventories, capital spending, discretionary expenses, and so on. So unemployment conditions are less of an obstacle to investing at this point in the cycle. However, economists fear that recent job and wealth losses will lead to further consumer weakness and accordingly a further downward spiral. Government subsidies and acquisition of shares at anchor and distressed businesses may boost trust into the market and excite growth insecurities, loan credit, and business financing. It remains to be seen if Government induction into the market is sustained, given the heavy debts most governments carry, to bring back a sustained growing economy, or if it is a single one off boost followed by decay. The global economy will not recover at the same speed as the financial industry. Dominated by the consumer spending, the global economy may only recover as soon as consumer confidence is back on track and spending becomes much relaxed.

Managing wealth in volatile markets is not straightforward. We see opportunities for investors of diverse risk profiles to try to fortify and grow their portfolios. Appropriate opportunistic acquisitions or mergers should be selected as part of an established portfolio strategy, either as a first step in building a portfolio or as a way to reposition assets within a fully invested portfolio. Using these ideas to capitalize on the current environment doesn't change a fundamental conviction: Over the long term, the best approach to grow assets and achieve multigenerational goals is based on developing a durable portfolio. The portfolio should be centered on strategic asset allocation and a suitable risk threshold that will allow investors to tolerate exposure to potential losses in the short term and stay invested consistently through all market cycles.

History has proven that market dislocations can lead to new and valuable opportunities. As investors position the portfolios that they manage cautiously, they continue to identify a wide range of ideas of all risk tolerance levels. By mid 2008, the *price* of a company and therefore its share had reached, on average, a 15:1 ratio versus its *earnings*. In some countries and some industries the ratio was even higher. By mid-2009 the *price-to-earnings* ratio of a company has deteriorated to fully discount market risk of an average of 6:1. Therefore, convertibles, or company loans that could be converted to a company share at the wish of the lender, provide good value proposition as the risk grows less while the potential for reward is big. Commodities, including energy, metals such as gold and agricultural crops, have provided portfolios with diversification benefits; something that has become a greater investor focus after the 2008 crisis and equity disorder.

Commodities also have generally provided effective hedges and bypasses against upside inflation risks and a weaker US dollar. After the dramatic, broad-based sell-off in commodities of 2008, there are more attractive valuations across this asset class by mid-2009. Further, the outlook for longer-term supplies across a number of commodities is constrained, while a global recovery suggests improving demand. The combination of these factors, together with the potential drop of crops as a result of changing temperature belts suggests higher prices over the coming years. Although performance suffered across most asset classes in 2008, investing in a broad range of asset classes did prove beneficial in helping investors keep more of their wealth than if they had taken a less diversified approach. Putting overall performance in perspective, broad diversification did help investors to perform better, or lose less, than had they been exposed to more concentrated positions. When one asset class sharply underperforms; as equities did in 2008, assets that outperform help to limit a portfolio's overall losses. For example, in 2008 certain credit classes and commodities had positive returns.

30-day T-bill (January 1926 to March 2008, annualized)

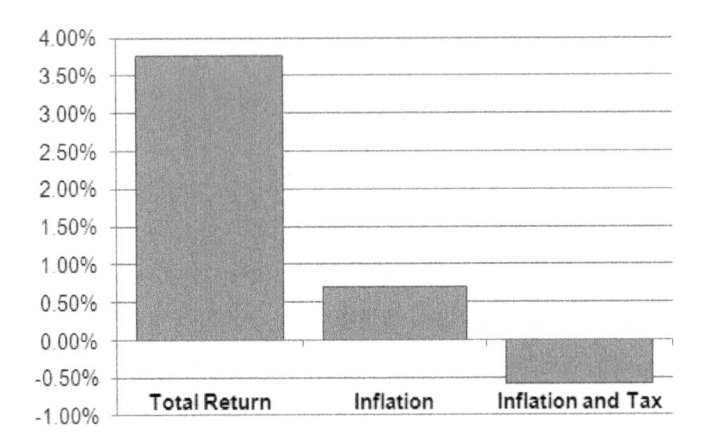

Data as of April 15, 2009
Source Ib boston Associates, Analysis assumed an effective aggregate Tax rate of 35%, the highest US Federal tax bracket for the full period

Figure 41- Cash returns after US federal taxes and inflation

And even though hedge funds posted losses, the asset class still outperformed equities by a healthy margin. That is why full diversification, including exposure to equities, fixed income, alternatives, commodities and cash, is still one of the most critical strategies in managing volatility and long-term returns. However, there remains the question of risk that governs every investment decision whatever the exposure is. When stocks shed more than 60 percent of their value in twelve months, as they did in 2008, investors' appetite for risk can justifiably evaporate. Staying calm and focused in the midst of turbulent markets can be difficult, especially when financial headlines fuel investor panic. It is no surprise that many people have opted for the safety of cash. While cash may indeed provide some degree of comfort and reassurance, for now, there can be unintended consequences on the ability to achieve long-term goals. Over time, cash concentrations can reduce the probability of maintaining purchasing power. The chart above shows the dramatic erosion of cash holdings due to inflation and taxes.

Agriculture, trade, manufacturing and financial services are the anchor industries upon which an economy flourishes. It took thousands of years to mature the agriculture industry, hundreds of years for trade and decades for manufacturing. It took less of a decade to bring a financial revolution. If history is a witness of anything, it is the cyclic pattern that governs our lives, relations, businesses. For instance, will we live through a new agriculture transformation? Climate change has started slow and is growing fast. It had, recently, compelled some forty countries to start saving grains of major crops in the fear of severe weather conditions that could lead to extension of some current plants' species. If environmental conditions change abruptly, an immediate loss of crops will occur. One may argue that the world storage of grains could cover our needs for a year or two. There is a danger of losing stored goods due to floods, hurricanes, fires and Earthquakes. There is a danger of losing utilities, infrastructures and transportation which otherwise work in harmony to secure transfer of grains from source locations to target markets.

Assuming all of the above could be managed to sustain supply versus demand, what will become of future crops? As the temperature and precipitation map of the Earth changes upon relocation of the ice caps, regions that are green in 2009 may not keep their green status anymore. Likewise, barren desert regions of 2009 could become lush and appropriate for sustained agriculture. One may imagine migrating agriculture machinery and skilled farmers to the new green regions is a matter of weeks if not months and business would resume as usual.

While people living in the desert regions are accustomed to continuous travelling and relocation wherever the wealth is and have, therefore, loyalty to the head of the tribe, people living in lush regions have intimate loyalty to the land where they live and may find it hard to relocate. But, the most important limitation to a smooth transformation of agriculture is governments controlling borders. Bureaucracy developed over decades to serve and protect the community of any one land is not ready for such sudden and abrupt change. The danger of war over fertile areas and water may persist if governments do not act quickly and collaborate to build a balanced win-win relationship. It would be prudent, therefore, to strike cross-border "Public- Private Partnerships", also known as PPPs, or "Climate Partnership", between the governments whose lands are currently deserts but soon to turn into fertile and lush and private businesses that have the liquidity and access to agriculture experience but, due to the current economic crisis, still have no vision of where or when to right-invest.

Infrastructure can be defined as the basic physical and organizational structures needed for the operation of a society or enterprise such as electricity grids, networks of water , sewage, telecommunication, or the services and facilities necessary for an economy to function, including roads, tunnels, bridges, railways, waterways, seaports, airports and so on. Governance of an organization or state that includes its bylaws, policies, processes and management system, is considered a basic infrastructure for the entity to function in a sustained manner. The health of an economy is directly related to how well its infrastructure is built, maintained, expanded and run. The supply chain or flow of material from one end (producer) to another end (consumer), for instance, becomes highly dependent on the availability, reliability and serviceability of the different layers of the infrastructure. It is common practice that redundancy is made integral to an infrastructure design. When an accident happens, an alternate route to sustain the service is made available, sometimes on full mode and sometimes on degraded mode of service.

While some infrastructure owners invest in setting up a duplicate capacity that is only deployed in case of disruption, others would go into mutual disaster/recovery agreements with other infrastructure owners of same trade. A good strategy would be to expand and ensure that such agreements are executable and not only well defined and confined on paper on the shelf. It is equally important to revisit the infrastructure development projects under construction. For example, the Great Rift Valley extends from Lebanon in the north to Mozambique in the south, and constitutes one of two distinct physiographic provinces of the East African Highlands physiographic division. The northernmost part of the rift, called the Dead Sea Transform fault (DST) system, is the major tectonic feature controlling the strati graphic and structural evolution of the region

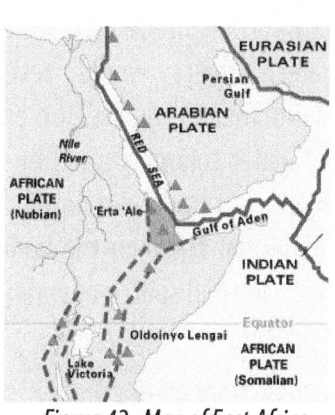

Figure 42- Map of East Africa showing line of tectonic shift

since the Miocene Age. A million years ago, a major Earthquake created the Arabian-African Rift. The Dead Sea sank deep into the valley and was deprived of its natural outflow to the sea. Today, the Dead Sea is the lowest point on Earth with its surface at 400 m (1,312 ft) below sea level. Fresh water flowing downstream through the Jordan River empties into the terminal lake. Having no exit point, the Dead Sea water evaporates, causing salts to accumulate in the lake and in its sediments. As a result, the Dead Sea's salt concentration is about 33 percent, compared to 3 percent in the Mediterranean. In the 1930s, the inflow of freshwater equaled the rate of evaporation, with the Jordan River emptying some 1.3 billion cubic meters (317 billion gallons) of clean water per year.

Today, the inflow is only one third due to national water projects on both sides of the Jordan River that have diverted freshwater upstream. As the rate of inflow from the Jordan River has decreased, so has the level of the Dead Sea. The lake's high rate of evaporation has also contributed to its declining level. The Dead Sea drying up, therefore, brings severe negative consequences on the ecosystem, industry and wildlife in the area. There have been several proposals for a canal to transport Mediterranean Sea or Red Sea water to the Dead Sea. Such a water project would reverse the negative impacts on the environment—that is, the erosion of the shoreline and disruption of the water column caused by declining water levels. The canal would create new trade and development opportunities by using the height differential between the bodies of water to generate hydropower, a much-needed source of water for domestic, agricultural and industrial purposes.

However, it would be prudent to assess the possibility of major Earth tremors and Earthquakes as discussed earlier in the book. What if, in a couple of years, such a canal is naturally entrenched! Land development and improvement does include infrastructure. Urban planning helps to optimize the cost and improve the speed of delivery of necessary civic and rural infrastructure while keeping the functional intention of the development intact. As a result of the sudden melting of the ice caps, the rising sea level is unavoidable, as are tidal waves that accompany the sudden discharge of molten ice into the ocean. It became apparent that future urban planning codes should minimize coastal cities. Instead, cities are better developed at least 15-20 km (9-12 miles) away from the coastal line or at 100 to 200 m (330 to 660 ft) altitude from the sea level.

Energy sources span from fossil, hydraulics, and nuclear to the latest sources of renewable energy. Earthquakes can hardly change the position of a coal mine. However, depending on the severity of the quake, an oil basin may leak its reserve if sediments cracks are deeply rooted and tunneled to other empty cavities and basins. The map detailing changes in climate and temperature would lead to a new pattern of water distribution. Hydraulic power plants erected on the path of abundant and rich water rivers may cease to operate if water flow dries out. Careful relocation of such plants could resolve the issue, but this would consume time and resources, as would, erecting new power grids to carry the resultant electrical current from the new locations. Renewable energy plants based on wind turbines, thermo or photovoltaic panels must follow the new meteorological map of the planet. Relocation could be carefully carried out, but in the absence of profitable returns on investment of renewable energy technologies at the current moment, such relocation may not come as a priority. It seems that nuclear power plants are the least impacted by a changed meteorological map. The design of such power plants, in most cases, has already taken into consideration Earth tremor and Earthquakes. However, safe shutdown procedures should be reexamined for severe, repetitive and prolonged tremors.

When we look at the Sun casually, it appears to be a uniform, unchanging star. Scientists and engineers have a very different perspective. To them, the Sun is a dynamic, chaotic, and poorly understood caldron of thermonuclear forces, one that can spit out fierce bursts of radiation at any time[24]. And when Earth lies in the path of that blast, the flare can play havoc with power grids, disrupt radio communications, and disturb or disable satellites. Fifty years into the space age, Earth has avoided the worst the Sun can deliver, so far. But with the Sun entering a period of increased activity, more frequent solar flares could be headed our way. With the weakening of the magnetosphere, are satellites and power grids ready to cope up with the increased rate of solar flares mass energy?

"Solar flares are a product of the Sun's complex chemistry", says Haimin Wang, director of the Space Weather Research Lab at the New Jersey Institute of Technology in Newark. The internal processes of the Sun create whirlpools of magnetic force, which slightly lower the surface temperature of the Sun, causing what we see as Sunspots. As referred earlier, the stronger the magnetic field the slower solar mass particles can oscillate at the thermosphere layer and accordingly less radiation energy and temperature is observed. Every eleven years, the Sun flips its magnetic north and south. This swap churns up an increased number of Sunspots, with the next volatile peak due in 2012. Magnetic energy can build up on the solar surface and then suddenly be released in a massive burst. This flare, a wave of particles moving near the speed of light, arrives at Earth shortly after the light itself, which takes about eight minutes to cross the 150 million km (93 million miles). In other words, sky-watchers won't know there has been a solar flare until shortly before the radiation arrives. The burst is largely cushioned by Earth's magnetosphere. Because of magnetosphere padding, the burst poses little risk to people on the ground. However, airplanes at high cruising altitude and spacecraft are

---

[24] http://features.csmonitor.com/innovation/2009/05/05/solar-storms-ahead-is-Earth-prepared/ Solar storms ahead: Is Earth prepared?

much more vulnerable with the weakening of the magnetic field. Although is based on a theory that the magnetic field that protects the Earth is generated at the Outer Core, which I am not in agreement with as explained and manifested earlier; but the film; *The Core* offers a speculative view of this situation. In the film, a severe electromechanical failure takes place. Equipment is broken and planes are crashed. In an analogical view, the Sun flare can zap electronics on satellites, which lie entirely in vulnerable magnetic zone outside the Earth's atmosphere. While this initial burst can be dangerous, there's a much slower wave of energy – a coronal mass ejection (CME)—that can seriously disrupt electronics. Not every flare produces a CME, and they often occur when no flare is present. When they do spew out, CMEs send strong waves of protons to the Earth. The most visible signs of this are the colourful northern and southern lights. But, a CME can also have more serious consequences. For one, it can cause electrical transformers to trip or fail, which can lead to widespread power outages. A particularly powerful CME storm hit Earth in 1921, before electricity played as big a role in daily life as it does now. If a burst of similar magnitude hit today, it would interrupt power for as many as 130 million people, (National-Acadmy-of-Sciences, 2009). In 1989, a geomagnetic storm knocked out power to 6 million people in Quebec. CMEs also can cause the Earth's atmosphere to expand temporarily. This can cause low-orbit satellites, such as the constellation of Global Positioning System markers used for navigation, to drag in the denser air. Combined with changes in the transmission of radio waves caused by CMEs, this can lead to errors in positioning. The magnetic field also can induce glitches or even damage satellites. Ron Mahmot, who manages the Satellite Operations Control Center for the National Oceanographic and Atmospheric Administration (NOAA), isn't particularly concerned about potential damage from solar activity.

NOAA has had only one satellite damaged by a solar flare. The 1994 incident merely shortened the life of the orbital weather-watcher; it did not totally disable it. And for the most critical NOAA satellites, those that monitor the United States from geosynchronous orbit, there is an on-orbit spare in place, with more still to launch. Dr. Guhathakurta of NASA is less certain: "In the past, satellites were built with much greater integrity", she says. "Now we're putting up so many satellites, and the electronics are changing. I don't believe the electronics are as well tested for radiation as in the past". Part of the problem, she says, is that we don't know what the most powerful solar flare might look like. And even if 2012 brings a mild solar peak in terms of quantity, the power of the flares is not directly associated with their number. If an orbiting planet that follows an elliptic path around two suns, as demonstrated in chapter three, circulates our Sun at very high speed, the magnetic pull of such a planet may not only increase the number of flares, but also increase their magnitude and outreach. The bulged weaker magnetosphere would bring less of a shield to protect the Earth from the charged particles of the Sun.

As we discussed earlier, magnetism is one of the forces in which materials through monopole photons exert attractive and repulsive force or momentums on other materials. Some well-known materials that exhibit easily detectable magnetic properties are nickel, iron, cobalt, gadolinium and their alloys. These substances, when charged, are known as magnets. All materials are influenced to greater or lesser degree by the presence of a magnetic field. Substances which are negligibly affected by magnetic fields are known as non-magnetic substances. They include copper, aluminum, water, and gases. Most generators and motors have magnetic components. The interaction of magnetic force lines and electric current produces motion, as in the case of a motor. Equally, the interaction of magnetic force lines and motion produces electric current in the body in motion, as in the case of a generator. Motors and Generators use ferromagnets, which keep their magnetization in the absence of an electric field. Motors and Generators also use electromagnets that acquire their magnetization only when electricity flows in a surrounding coil. Neighboring pairs of electron do spin to point in opposite directions. This property is not contradictory, because in the optimal geometrical arrangement, there is more magnetic momentum from the sub-configuration of electrons which point in one direction, than from the sub-configuration of electrons which points in the opposite direction. Thus, a net magnetic force in the direction of the geometrically prevailing electron' spin is produced. We currently know that every ferromagnetic substance has its own individual temperature, called the Curie temperature, or Curie point, above which it loses its ferromagnetic properties. Fortunately, this temperature is forty times the room temperature at 1 bar pressure, otherwise, ferromagnets would lose their magnetic properties every now and then and motors and generators would become highly unreliable. The magnetosphere is over bulged and weakened as its force lines are re-routed to meet up with the magnetic force lines of the *mPlanet*. The swarm of charged particles of the solar storms, such as positrons, and protons, that are in plasma state, will find it easy to invade the magnetosphere. It will bathe all metals including ferromagnets, hunt the pairs of free electrons and destroy the spin imbalance of the electron pair configuration. Lesser pairs of free electrons lead to lesser magnetic force and failure for the metal to act as ferromagnets. Equipment that uses ferromagnets and electromagnets to function will

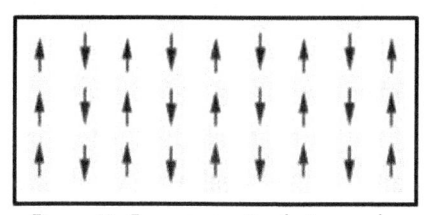

Figure 43- Ferromagnetic electron order

cease to operate. I suggest shielding the electronic boards, motor magnets, and generators magnets by non-magnetic substance that shields such plasma-state protons and positrons away from the magnets. As scientists gather more information, satellite companies have a few tricks to protect their space-bound electronics. It is hard to determine the effectiveness of shielding, mostly because experts don't fully know what the Sun is capable of spewing out. Another satellite defense would be "safe modes", in which administrators turn off most of the on-board electronics to protect them from going haywire. This can save the satellite but would interrupt whatever services it offered. Safe modes also require some kind of early warning; the technology for this is nascent at best.

Planet life is very vulnerable to changes in diet, habitat temperature and reproduction processes such as following certain migratory paths at certain times of the year. If the food chain is broken due to extinction of some species, and if no alternative is made available, the whole life chain on top could face elimination. Climate change may drive a quarter of land animals and plants extinct, according to a major new study (Thomas & Hannah, 2004) While the study attributes the cause to rising planet temperature due to greenhouse gas emissions—which, according to my earlier analyses, could not be the prime cause—there will be drastic and swift changes in the habitats of most species. The largest consortium of scientists ever to apply themselves to this problem studied six biologically diverse regions around the world representing 20 percent of the planet's land area. They projected the future distributions of 1103 plants, mammals, birds, reptiles, butterflies and other invertebrates. The study employed computer models to simulate the ways species' ranges are expected to move in response to changing temperatures and climatic conditions. Using data supplied by the Intergovernmental Panel on Climate Change in the USA, the scientists considered three different climate change scenarios: minimum, mid-range and maximum expected climate change over different regions of Earth. They also considered the ability of animals and plants to move to new areas, using two alternatives: one in which species could not spread at all, and the other assuming "unlimited" or successful movement.

The study found that 15 to 37 percent of all species in the regions considered could be driven extinct from the climate change that is likely to occur between now and 2050. However, the study should be re-visited in connection with abrupt rather than gradual change. The scientists believe that extinctions due to climate change are also likely to occur elsewhere. "If the projections can be extrapolated globally, and to other groups of land animals and plants, our analyses suggest that well over a million species could be threatened with extinction as a result of climate change", said lead author Chris Thomas a biologist at the University of Leeds, England. So what could be done? We can reflect on the lessons of Noah's Ark We have also heard of the efforts of some forty countries of the western hemisphere to store samples of most if not all kinds of grains currently available. Could a similar effort be made to include all plants and animals? Fish and mammals at sea will be able to relocate more freely. However, they will suffer severe oceanic volcanoes that rip the Crust open, spewing lava at sea. They will get stranded in pockets of water or dead-ends as well as shallow creeks if they rely on Earth's magnetic force lines, hardly predictable, to guide them navigate at sea.

A study, at the University of Florida (DeSantis, 2009), shows that mammals change their dietary niches based on climate-driven environmental changes. The study contradicts a common assumption that species maintain their niches despite global warming. Led by Florida Museum of Natural History vertebrate paleontologist Larisa DeSantis, researchers examined fossil teeth from mammals at two sites representing different climates in Florida: a glacial period about 1.9 million years ago and a warmer, interglacial period about 1.3 million years ago. The researchers found that interglacial warming resulted in dramatic changes to the diets of animal groups at both sites. Co-author Robert Feranec, curator of vertebrate paleontology at the New York State Museum, said that scientists cannot predict what species will do based on their current ecology. "The study definitively shows that climate change has an effect on ecosystems and mammals, and that the responses are much more complex than we might think", explained Feranec. "The two sites in the study, both on Florida's Gulf Coast, have been excavated quite extensively", said DeSantis. During glacial periods, lower sea levels nearly doubled Florida's width, compared with interglacial periods. But because of Florida's low latitude, no ice sheets were present during the glacial period.

Despite the lack of glaciers in Florida, the two sites show that dramatic ecological changes occurred between the two periods. Both sites include some of the same animal groups. The research examined carbon and oxygen isotopes within tooth enamel to understand the diets of medium to large mammals, including pronghorn, deer, llamas, peccaries, tapirs, horses, mastodons, mammoths and gomphotheres, a group of extinct elephant-like animals. Differences in how plants photosynthesize give them distinct carbon isotope ratios. For example, trees and shrubs process carbon dioxide differently from the way warm-season grasses process this gas, resulting in different carbon isotope ratios. These differences are incorporated in mammalian tooth enamel, allowing scientists to determine the diets of fossil mammals.

Lower ratio values suggest a browsing diet (trees and shrubs) while a higher ratio suggests a grazing diet (grasses).

Animals at the glacial site were predominantly dieting on trees and shrubs, while some of those same animals at the warmer interglacial site became mixed feeders that also grazed on grasses. Increased consumption of grasses by mixed feeders and elephant-like mammals indicates that Florida's grasslands likely expanded during interglacial periods. To find these signatures, researchers run samples of tooth enamel through a mass spectrometer. "When people are modeling future mammal distributions, they're assuming that the niches of mammals today are going to be the same in the future", said DeSantis. "That's a huge assumption".

Civil order has been developed over centuries of elaboration and speculative approaches. The crafting of laws, policies, regulatory and controls led to delivery of sound services on the government, public and private levels. The eventual application of information technology enabled societies, communities and companies to prosper through transparent, spontaneous, correct and effective services. The Internet seemed to have kept a good and growing memory of knowledge to support more effective decision making, collaboration and communication. All that is required, for a police officer who wants an assurance that the correct procedure is being applied, is to type an inquiry at a networked computer. The officer would also go through online criminal records to limit the search for suspects and become more responsive in closing a criminal case. What if the officer, as it was half a century ago, had to look into text books, manuals and piles of papers and mug books of suspected persons? Would he or she be trained to do that efficiently (quickly) and effectively (exhausting all possibilities that are pertinent to the case in hand)? Probably not!

Communities' expectations would have to be set realistically low. It would be some time before recovery and reestablishment of computer storage, processing and networking capacities occurred. It is prudent to set and test strategies for data protection and disaster recovery procedures. Scenarios of manual operation must be laid down, and service officers must go through training and a learning curve of operation and probably communicating using disciplines that have otherwise been extinct for more than 100 years.

A national security team will typically include any one country Department of Defense, the intelligence community and allies. Industrial security will have to include critical infrastructures as well as government and nongovernmental organization services. Civil security would be typically led by the Department of Homeland Security, working hard to maximize law enforcement. The network of those three teams must be continuous, shared and prepared to protect communities from key threats and dangerous events. The human element remains the key factor to ensure that i) legal codes are adhered to, ii) public infrastructures are resilient and redundant through either additional resources or bilateral agreements, iii) the economy is prepared to handle shortages and cut off routes, and much more. It is crucial to revisit the immigration control policies in anticipation of potential and massive human migrations. After the storm is over and a new ecological map of the planet emerges, borders may be used to delineate only differences in language and deep-rooted traditions.

Lead country projects, is a simple concept where each country of the world is assigned a leading responsibility or a set of leading responsibilities to manage the preparation and transformation during Earth Changes. As in every successful project, a timely, quantitative, and qualitative vision is defined and agreed upon by all involved. There follows, the organization of every project, wherein a lead country has a set of responsibilities and led countries have complementary and supporting ones. The utilization of information technology is a must to provide instantaneous data and information sharing round the clock and across the planet. While such projects assemble and embrace the expertise, knowledge and wisdom of project managers, industry experts and those who maintain community relations, there should be a pool of shared services and competence in finance, human resources, legislation, logistics, and technology.

I believe that the human race draws closer at times of calamity. In the likely event that wars will cease to exist for a good period of time, the United Nations, could, reinvent itself for a period of time, as the Policy Maker, stating the rules and guidelines that will aid the transformation. Nongovernmental organizations (NGOs) could play facilitator roles in such global and cross country projects. The United Nations climate change conference in Poznań concluded in December 2008 with a clear commitment from governments to shift into full negotiating mode in order to shape an ambitious and effective international response to climate change, to be agreed upon in Copenhagen at the end of 2009; which did not turn out. The core of all discussions and negotiations must change from a narrow view to set rules and quota to control carbon dioxide emission. As proven earlier in this book, carbon dioxide emission is innocent from climate change. Instead and based on the actual cause, the effort should be exerted to

Figure 44- Organization Role

draw a holistic time view on Earth changes and to lay down strategies, projects and roles to minimize the impact of such changes.

New national and international policies, changing consumer preferences and emerging markets in response to climate change are stimulating new trade and investment opportunities. Such opportunities are being picked up by big business, but small to medium size enterprises are structurally disadvantaged and poorly prepared to seize the new opportunities. However, the main issue is that most investment opportunities are focusing on low-carbon solutions and the creation of green collars jobs. A just call but far from right as '0' carbon emission will not stop the Earth Changes that is caused by purely natural and cyclical phenomena. The urgency of the matter calls for humankind to revisit its priorities and direct the funds and resources to the right cause: setting a simple and achievable set of strategic actions and keeping societies and communities aware and supportive of the progress. There should be no independent gain for any one country or any single community or person. A global problem mandates a global solution. Let us take an example and ask what could, a global solution to a world-wide famine scenario be? The World Trade Organization is working to close the Doha round of trade negotiations so that a consistent trade policy is applied across the countries of the world. The round should reduce subsidies, and would lower tariff walls in developed and developing countries. For the food sector, this means bringing food closer to the poor. The discussion may take forever as it touches on hundreds of countries, thousands of categories of traded products and services and millions of permutations of components country of origin. I believe that climate change impacted goods and services should singled out of such prolonged discussion and be given an absolute top priority in the round of trade negotiations to conclude fast and get adequate flow and distributed storages of necessary crops enacted ahead of time. We can never be sure of roads and transportation availability when the time comes.

Gain should be collective and for all. We live on a single ship called Earth. A hole in one corner will cause it to sink, and those on board will share same destiny no matter how high they stood. We should understand the past, study the present and be prepared for the future. At the tail of every end there is always a new beginning.

Bibliography

- Abu-Salieh, S. A. (1999). Muslims' Genitalia in the Hands of the Clergy. *Fifth International Symposium on Sexual Mutilations* (p. 141). Oxford: Kluwer Academic/ Plenum Publishers.
- Australian-Antarctic-Division. (2002). *South Magnetic Pole.* Commonwealth of Australia.
- Boothroyd, A. (2009). *Magnetic monopoles: 70 years from prediction to observation.* Institut-Laue-Langevin.
- Brown, G. C., & Mussett, A. E. (1981). *The Inaccessible Earth.* Taylor & Francis.
- C.S. Sovers, O.J Archinal, P. Charlot. (2000). Annual Report 2000. *International Earth Rotation Service* .
- Canada-Natural-Resources. (2005). *Geomagnetism, North Magnetic Pole.*
- Cruickshank, D. (Director). (2007). *Around the World in 80 Treasures* [Motion Picture].
- David C. Catling, K. J. (2001). Biogenic Methane, Hydrogen Escape, and the Irreversible Oxidation of Early Earth. *Science 293* , pp. 839–843.
- DeSantis, L. (2009, June 2). *UF study finds that ancient mammals shifted diets as climate changed.* Retrieved from Public of Library Science: http://www.eurekalert.org/pub_releases/2009.../plos-usf052909.php
- Division, A. A. Common Wealth of Australia.
- Duennebier, F. (1999). Pacific Plate Motion. *University of Hawaii* .
- Egypt, S. I. (2006, November 14). *Ancient Egypt.* Retrieved from Tour Egypt: www.touregypt.net/suezcanal.htm
- Gilbert, A. (2007). *The End of Time- The Mayan Prophecies Revisited.* Mainstream Publishing.
- GIRIJA RAJARAM, T. A. (2002). *Rapid decrease in total magnetic field F at Antarctic stations.* (C. M. Indian

Institute of Geomagnetism, Ed.) Retrieved 2009, from Antarctic Science 14 (l), 61-68: http://journals.cambridge.org/download.php?file=%2FANS%2FANS14_01%2FS0954102002000585a.pdf&code=813e17119a1fce1a5eecd461067b11c3
- Hancock, G. (1995). *Fingerprits of the Gods.* Three Rivers Press.
- Hancock, G. (1998). *The Mars Mystery, The Secret Connection linking Earth's ancient civilization and the Red Planet.* Three Rivers Press.
- Hawking, S. (1991). *Quest for a Theory of Everything.* Bantam Books.
- IERS, I. E. (2000).
- Lazar, S., Treadway, M., & Chakrapami, S. (2006, January 23). *Science/ Research.* Retrieved from Harvard University Gazette: http://www.news.harvard.edu/gazette/daily/2006/01/23-meditation.html
- Lloyd, S. (2005). *Programming the Universe- A quantum Computer Scientist takes on the Cosmos.* London: Vintage Books.
- Macchi, M., & Bruce, J. (2004). *Human pineal physiology and functional significance of melatonin.* Front Neuroendocrinol .
- Marrs, J. (1997). *Alien Agenda- Investigating the Extraterresterial presence among us.* Harper Paperbacks.
- McMoneagle, J. (1998). *The Ultimate Time Machine- A Remote Viewer's Perception of Time and Predictions for the New Millennium.* Hampton Roads Publishing Company Inc.
- Milbert, G., & Smith, D. (2007). Converting GPS Height into NAVD88 Elevation with the GEOID96 Geoid Height Model. *National Geodetic Survey, NOAA* .
- Moran, J. (2005). *Weather.* NASA/World Book, Inc.

- Mozes, R. B. (2007). *First Ever*. Retrieved from Traditional Circumcision: http://www.britpro.com/default.asp?p=first
- National-Acadmy-of-Sciences. (2009). *Severe Space Weather Events-Understanding Societal and Economic Impacts*. National Academies Press.
- Newberg, A. (2002, March 1). *Science and Technology*. Retrieved from BBC News: http://news.bbc.co.uk/2/hi/science/nature/1847442.stm
- Ng, S. K. (2002, 10 17). *Magnetic monopole is photon*. Retrieved from Research Gate- Scientific Network: https://www.researchgate.net/publication/2054444_Magnetic_monopole_is_photon
- Penrose, R., & Gardner, M. (2002). *The Emperor's New Mind: Concerning Computers, Minds and Physics (Popular Science)*. Oxford: Oxford University Press.
- Philips, G. (1998). *Act of God Tutankhamen, Moses and the Myth of Atlantis*. HB Sidgwick & Jackson, PB Pan .
- Pidwirny, M. (2006). *Fundamentals of Physical Geography*. PhysicalGeography.net. http://www.physicalgeography.net/fundamentals/7h.html.
- Radford, B. (2007, July 21). *The Ten-Percent Myth*. Retrieved from Snopes.com: http://www.snopes.com/science/stat/10percnt.html
- Sanders, R. (2003). Radioactive potassium may be major heat source in Earth's core. *UC Berkeley News* .
- Schaefer, L. (1997). *In Search of Divine Reality; Science as a Source of Inspiration*. Fayetteville: University of Arkansas Press.
- Smith, G. E. (1910). Circumcision in Ancient Egypt. *British Medical Journa* , 294.
- Solomon, S., Plattner, G.-K., Knutti, R., & Friedlingstein, P. (2009). Irreversible climate change due to carbon dioxide emissions. *National Academy of Science* .
- Targ, R. (2004). *Limitless Mind- a guide to remote viewing and transformation of consciousness*. New World Library.

- Thomas, C., & Hannah, L. (2004, July 8). Science News. *Nature* .
- Thompson, A. (2008, 12 16). Leaks Found in Earth's Protective Magnetic Shield. *Space.com. Imaginova Corp* .
- Tilak, B. G. (1903). *The Arctic Home in the Vedas.* Arktos Media Limited.
- University-of-California-Los-Angeles. (2009, May 13). *Science News.* Retrieved from Science Daily: http://www.sciencedaily.com/releases/2009/05/090512134655.htm
- Williams, D. R. (2004). *Earth Fact Sheet.* NASA. http://nssdc.gsfc.nasa.gov/planetary/factsheet/earthfact.html.
- Wilson, I. (1985). *Exodus Enigma.* Weidenfeld & Nicolson.

Tarek S. Niazi holds a Bachelor degree and a Masters degree in Electrical Engineering and Computer Science respectively. He worked at IBM for twenty-five years in various positions. As an entrepreneur, he established and ran investment, consulting, and project integration services companies. Niazi's interests run the gamut from geology to history, astronomy to theology, and most of the sciences, including physics. His global travels have given him a deep appreciation for the mosaic of life and culture on Earth. More updates and videos may be found on http://www.planet-earth-2017.com